AF214462

Defects in Organic Semiconductors and Devices

Defects in Organic Semiconductors and Devices

Thien-Phap Nguyen

iSTE

WILEY

First published 2023 in Great Britain and the United States by ISTE Ltd and John Wiley & Sons, Inc.

ISTE Ltd
27-37 St George's Road
London SW19 4EU
UK

www.iste.co.uk

John Wiley & Sons, Inc.
111 River Street
Hoboken, NJ 07030
USA

www.wiley.com

Library of Congress Control Number: 2023937986

British Library Cataloguing-in-Publication Data
A CIP record for this book is available from the British Library
ISBN 978-1-78630-926-6

Contents

Abbreviations

General abbreviations

AFM	Atomic Force Microscopy
BHJ	Bulk Heterojunction
CELIV	Charge Carrier Extraction by Linearly Increasing Voltage
CPE	Constant Phase Element
CTC	Charge Transfer Complex
DLCP	Drive-Level Capacitance Profiling
DLOS	Deep-Level Optical Spectroscopy
DLTS	Deep-Level Transient Spectroscopy
DOS	Density of States
EA	Electron Affinity
EBL	Electron Blocking Layer
EDX	Energy-Dispersive X-Ray Spectroscopy
EIL	Electron Injection Layer
EML	Emitting Layer

ENDOR	Electron Nuclear Double Resonance
EQE	External Quantum Efficiency
ESR–EPR	Electron Spin Resonance–Electron Paramagnetic Resonance
ETL	Electron Transport Layer
FL	Fluorescence
FLIM	Fluorescence Lifetime Imaging Microscopy
GDM	Gaussian Disorder Model
HBL	Hole-Blocking Layer
HIL	Hole Injection Layer
HOMO	Highest Occupied Molecular Orbital
HTL	Hole Transport Layer
IE	Ionization Energy
IQE	Internal Quantum Efficiency
IS	Impedance Spectroscopy
ITC	Ionic Thermo-Current
KPFM	Kelvin Probe Force Microscopy
LESR	Light-Induced Electron Spin Resonance
LUMO	Lowest Unoccupied Molecular Orbital
NFA	Non-Fullerene Acceptors
NREL	National Renewable Energy Laboratory
OFET	Organic Field-Effect Transistor
OLED	Organic Light-Emitting Diode
OLET	Organic Light-Emitting Transistor

OPV–OSC	Organic Photovoltaic–Organic Solar Cell
OTR	Oxygen Transmission Rate
PCE	Power Conversion Efficiency
PDS	Photothermal Deflection Spectroscopy
PL	Photoluminescence
PLQY	Photoluminescence Quantum Yield
QD	Quantum Dot
RISC	Reverse Intersystem Crossing
SCLC	Space Charge-Limited Current
SQUID	Superconducting Quantum Interference Device
SSPL	Steady-State Photoluminescence
STM	Scanning Tunneling Microscopy
TADF	Thermally Activated Delayed Fluorescence
TAS	Transient Absorption Spectroscopy
TEM	Transmission Electron Microscopy
TOF	Time of Flight
TOF-SIMS	Time-of-Flight Secondary Ion Mass Spectroscopy
TRPL	Time-Resolved Photoluminescence
TSC	Thermally Stimulated Current
TSL	Thermally Stimulated Luminescence
TSPC	Thermally Stimulated Polarization Current
UPS	Ultraviolet Photoemission Spectroscopy
VOFET	Vertical Organic Field-Effect Transistor

WAXS	Wide-Angle X-Ray Scattering
WOLED	White Organic Light-Emitting Diode
WVTR	Water Vapor Transmission Rate
XPS	X-Ray Photoemission Spectroscopy

Chemical materials

1-NaphDATA	4,4′,4″-tris(N-2-naphthyl)-N-phenylamino-triphenylamine
4T	α-quaterthiophene
6T	α-sexithiophene
Alq3	tris(8-hydroxyquinolinato) aluminum
BCF	tris(penta fluorophenyl) borane
BDT	benzodithiophene
Bphen	4,7-diphenyl-1,10-phenanthroline
BT-CIC	(4,4,10,10-tetrakis(4-hexylphenyl)-5,11-(2-ethylhexyloxy)-4,10-dihydrodithienyl[1,2-*b*:4,5*b*’] benzodithiophene-2,8-diyl)bis(2-(3-oxo-2,3-dihydroinden-5,6-dichloro-1-ylidene)malononitrile)
BTA3	benzotriazole
BTP-eC9-2Cl	2,2′-[[12,13-bis(2-butyloctyl)-12,13-dihydro-3,9-dinonylbisthieno [2″,3″:4′,5′] thieno[2′,3′:4,5] pyrrolo[3,2-e:2′,3′-g][1–3] benzothiadiazole-2,10-diyl]bis[methylidyne(5,6-chloro-3-oxo-1H-indene-2,1(3H)-diylidene)]] bis[propanedinitrile]
CBP	4,4′-bis(N-carbazolyl)-1,1′-biphenyl
CdSe	cadmium selenide
$CH_3NH_3PbBr_3$	methylammonium lead tribromide

$CH_3NH_3PbI_3$ (MAPI)	methylammonium lead triiodide
CuOx	copper oxide
CuPc	copper phthalocyanine
CuSCN	copper(I) thiocyanate
DH4T	dihexyl-quaterthiophene
F4-TCNQ	tetrafluoro tetracyanoquinodimethane
$FAPbI_3$	formamidine lead triiodide
FIrpic	iridium(III)bis(4,6-(difluorophenyl)pyridinato-N,C2′)picolinate
HATCN	hexa-azatri-phenylene-hexanitrile
HMDS	hexamethyl-disilazane
IDTBR	(5Z,5′Z)-5,5′-((7,7′-(4,4,9,9-tetraoctyl-4,9-dihydro-s-indaceno[1,2-b:5,6-b′]dithiophene-2,7-diyl) bis(benzo[c][1,2,5]thiadiazole-7,4-diyl))bis(methanylylidene)) bis(3-ethyl-2-thioxothiazolidin-4-one)
Ir(ppy)3	fac-tris-(2-phenylpyridine)iridium(III)
ITIC ($C_{94}H_{82}N_4O_2S_4$)	3,9-bis(2-methylene-(3-(1,1-dicyanomethylene)-indanone))-5,5,11,11-tetrakis(4-hexylphenyl)-dithieno[2,3-d:2′,3′-d′]-s-indaceno[1,2-b:5,6-b′] dithiophene
LiF	lithium fluoride
MDMO-PPV	poly[2-methoxy-5-(3′,7′-dimethyloctyloxy)-p-phenylene vinylene]
MEH-PPV	poly[2-methoxy-5-(2′-ethylhexyloxy)-p-phenylene vinylene]
MeLPPP	methyl-substituted ladder-type poly para-phenylene

MeO-TPD	N,N′-diphenyl-N,N′-bis(3-methyl-phenyl)-[1,1′-biphenyl]-4,4′-diamine
MoO_3	molybdenum(VI) oxide
NPB	N,N′-diphenyl-N,N′-bis(1-naphthyl)-1,1′-biphenyl-4,4′-diamine
NPD	N,N′-di(1-naphthyl)-N,N′-diphenylbenzidin
NRS-PPV	poly[{2-[4-(3′,7′-dimethyloctyloxyphenyl)]}-co-{2-methoxy-5-(3′,7′-dimethyl octyloxy)}-1,4-phenylene vinylene]
OC_1C_{10}-PPV	poly[2-methoxy-5-(3′,7′-dimethyloctyloxy)-p-phenylene vinylene]
OC_8C_8	poly[p-(2,5-di(2-ethylhexyloxy)phenylenevinylene]
P3DDT	poly(3-dodecyl thiophene-2,5-diyl)
P3MeT	poly(3-methylthiophene)
PBD	2-(4-biphenylyl)-5-(4-tert-butyl-phenyl)-(1,3,4-oxadiazole)
PBDB-TF	poly[(2,6-(4,8-bis(5-(2-ethylhexyl)thiophen-2-yl)-benzo[1,2-b:4,5-b']dithiophene))-alt-(5,5-(1',3'-di-2-thienyl-5',7'- bis(2-ethylhexyl)benzo[1',2'-c:4',5'-c'] dithiophene-4,8-dione))]
PBQx	benzodithiophene-dithieno[3,2-f:2′,3-h]quinoxaline
Pc	pentacene
PCBM	[6,6]-phenyl-C61-butyric acid methyl ester
PCDA	10,12-pentacosadiynoic acid
PCDTBT	poly[N-9'-heptadecanyl-2,7-carbazol-alt-5,5-(4',7'-di-2-thienyl-2',1',3'-benzothiadiazol)]

PCNEPV	poly[oxa-1,4-phenylene-(1-cyano-1,2-vinylene)-(2-methoxy-5-(3',7'-dimethyloctyloxy)-1,4-phenylene)-1,2-(2-cyanovinylene)-1,4-phenylene]
PDI	perylene diimide
PEDOT:PSS	poly(3,4-ethylenedioxythiophene) polystyrene sulfonate
PET	polyethylene-terephthalate
PF-N-Ph	poly(9,9-dihexylfluorene-co-N,N-di(9,9-dihexyl-2-fluorenyl)-N-phenylamine)
PMPSi	polymethyl-phenylsilylene
PPP	poly(p-phenylene)
PPQ	phenyl-quinoxaline
PPV	poly(p-phenylene-vinylene)
PRA	1-phenyl-3-(p-diethylaminostyryl)-5-(p-diethylaminophenyl) pyrazoline
PTAA	poly(triarylamine)
PTB7	poly[[4,8-bis[(2-ethylhexyl)oxy]benzo[1,2-b:4,5-b′] dithiophene-2,6-diyl] [3-fluoro-2-[(2-ethylhexyl)carbonyl] thieno[3,4-b]thiophenediyl]]
PTCBI	3,4,9,10-perylenetetracarboxylic-bis-benzimidazole
PVK	poly(9-vinylcarbazole)
rrP3DDT	regio-regular poly(3-dodecyl thiophene-2,5-diyl)
Spiro-OMeTAD	(2,2′,7,7′-tetrakis(N,N-di-p-methoxy phenylamine)-9,9′-spirobifluorene)
SubPc	subphthalocyanine-chloride
TPD	N,N′-diphenyl-N,N′-bis(3-methylphenyl)(1,1′-biphenyl)-4,4′-diamin

TPQ	trisphenyl-quinoxaline
Y6 (BTB-4F)	2,2′-((2Z,2′Z)-((12,13-bis(2-ethylhexyl)-3,9-diundecyl-12,13dihydro[1,2,5]thiadiazolo[3,4e]thieno[2″,3″:4′,5′]thieno [2′,3′:4,5]pyrrolo[3,2-g]thieno[2′,3′:4,5]thieno[3,2-b]indole-2,10diyl)bis(methanylylidene))bis(5,6-difluoro-3-oxo-2,3-dihydro-1H-indene-2,1 diylidene))dimalononitrile
ZnPc	zinc phthalocyanine

Introduction

In a movie that was directed by Billy Wilder, one character declared "Well… nobody's perfect!" in order to excuse his partner's shortcomings, that is, in this situation, his way of life, which was substantially different from what he had portrayed. This statement may seem obvious, although no one can claim that they know everyone around them completely. On the other hand, in the field of materials and devices, it is well known that nothing is perfect due to the presence of defects within their structures and architectures. For instance, the density of defects in conventional semiconductors, such as silicon, is estimated to be higher than 10^{11} cm^{-3} at room temperature for intrinsic samples. As defects affect the quality and the properties of materials, and, consequently, the performance of devices using them, it is essential to control their density in order to ensure their reliability. Indeed, when we create, produce or acquire a material or device, we would like to have the best performance and the longest lifetime from its use. This requires careful control of not only the production process but also the characteristics of the materials used. In electronic devices, the physical properties of semiconductors are strongly dependent on the defect states, and it is essential to identify and understand their formations, their locations and to determine their densities in order to obtain reproducible materials with known and controlled defect parameters and to establish their reliability.

Conventional semiconductors such as silicon and germanium are crystalline solids, where the atoms form a periodic arrangement. This ordered structure provides highly interesting electrical and optical properties to the semiconductors that are used to build electronic components and devices. Since the invention of transistors, myriad applications have been achieved in the field of electronics, bringing great comfort to everyday life: TVs, lighting, computers and cell phones, to name but a few. However, as stated by Victor Hugo, "*On voit les qualités de loin et les défauts de près*" (we see qualities at a distance and defects at close range), despite their remarkable and numerous qualities, it was very quickly observed that many of the first electronic devices manufactured using conventional semiconductors malfunctioned, despite careful control of the processing. Through investigations of the defective parts, it was found and later proved that, irrespective of the particular devices, the nature of the materials played a primary role in the reliability of the electronic products, whose yield is closely linked to the presence of defects.

As perfect materials do not exist, the properties of conventional semiconductors are affected by defects which interrupt the crystalline pattern. Common types of defects include point defects (impurities, interstitials, vacancies, etc.), dislocations, and grain boundaries can be formed during the processing but can also be intentionally (doping) or unintentionally (contamination, degradation) incorporated in the prepared materials. In the doping process, impurities are intentionally introduced to materials in order to modify and control the conduction of the semiconductors by adding energy states in a band gap, which provide charge carriers to the conduction or valence bands. These defects have a beneficial effect on the electrical properties of the materials. In most other cases, defects have negative or detrimental effects on the properties and functionalities of materials by enhancing the disorder, impeding the charge transport and affecting the overall physical processes in the semiconductors. As defects are unavoidable, it is necessary to acquire accurate knowledge of their origin and their effects in order to efficiently control and eventually eliminate them. Investigations of defects in conventional semiconductors have been intensively developed with well-established and elaborated measurement techniques in order to determine the defect parameters in materials and devices, improving the knowledge of their origin and their effects on the performance of the devices studied. At the same time, diverse physical models on the material structure, energetic distribution and charge carrier kinetics have been proposed and successfully applied in order to elucidate defect measurement results in most conventional semiconductors.

Structurally speaking, organic semiconductors differ from conventional semiconductors. Since there is no defined orientation and order of molecules that make up the organic matter, they can be classified as amorphous materials. The lack of orientational order combined with the weak van der Waals bonding forces make the organic materials likely to form defects, which can be explained by the small amount of energy needed to displace the molecule from its equilibrium position. Indeed, similarly to inorganic semiconductors, impurities and structural defects such as point defects, dislocations and grain boundaries can be formed or introduced to the organic semiconductors during the synthesis and the processing of materials.

Due to their nature and chemical structure, they are also more sensitive than their inorganic counterparts to contact with environmental media. The structural changes due to the interactions between the organic material and the environment often lead to the formation of defects in the contact region. From this consideration, we can expect defects in organic materials to be investigated by applying similar methodology and techniques as in conventional semiconductors. To take the specific properties of the organic materials into account, further advanced measurement techniques and methodology approaches need to be used and developed, and the results obtained must be effectively analyzed and used.

This book aims to provide a comprehensive introduction on the defects and degradation of semiconductors used in electronic organic devices. It is organized as follows:

The first chapter is an overview of organic semiconductors and devices, in which fundamental notions of the materials and the main applications in organic electronics are presented.

Chapter 2 reviews the concept of defects in inorganic and organic semiconductors in relation to the notion of order/disorder, the density of states (DOS) and the localized states in the band gap.

In Chapter 3, the effects of defects on the electrical and optical properties of the organic materials are described. These properties are of primary importance for the operation of organic devices such as OLEDs and OPVs and also for defect measurement techniques.

Chapter 4 presents the main measurement techniques for determining the defect parameters in both organic and conventional semiconductors, which include paramagnetic resonance, optical and electrical techniques. The principle of the methods is described, the analysis of the results to extract the defect parameters (whenever possible) is explained and for applications, selected typical examples of defect measurement in organic devices from the literature are given.

Chapter 5 reports the results obtained from the defect measurements in different organic semiconductors and devices. Defects from the active layer, the transport layers, from the surface, and interface and surface, and from diffused impurities are detected in the organic devices studied.

Chapter 6 presents the correlation between defects and reliability in organic devices by studying the influence defects have on the efficiency, the lifetime and the degradation processes of devices such as OLEDs, OPVs and OFETs.

I would like to thank ISTE Ltd for publishing this book, and I hope that it will help readers to better understand some aspects of defects in organic semiconductors and devices.

1

Overview of Organic Semiconductors

Organic materials are defined as those that contain carbon–hydrogen (C–H) bonds. Other types of bonding of carbon atoms also exist, forming molecules within the materials. Contrarily to inorganic materials, in which atoms are bound by covalent or ionic bonds, the molecules are bonded by van der Waals forces. As a result, the materials are mechanically soft and flexible as compared to hard and rigid inorganic materials. This particular property can be an asset for the practical applications of organic devices but under various aspects also presents disadvantages to be overcome for their use.

The description of the structure of organic materials can be summarized as follows. The smallest molecular structure is a single unit called a *monomer*, whose weight can be determined from its chemical composition. An assembly of a defined number of monomers constitutes an *oligomer*, which is composed of monomers linked together either by carbon bonds (C–C) of the molecules or by linking units. The molecular weight of an oligomer is determined by the exact number of the constituting monomers. When the number of monomers increases, let us say more than 10 units, the size of the assembly of molecules becomes large and the assembly is called a *polymer chain*. The organic materials for electronics are classified into two different types: small molecules and polymers. *Small molecules* having a well-defined molecular weight include monomers, oligomers and dendrimers. The latter is a special assembly of monomers with a central core and symmetrically arranged branches with respect to the center. *Polymers* are composed of chains of indefinite length, and their molecular weight can be estimated but not accurately determined. As there is no periodical arrangement of molecules or polymer chains, the organic materials are globally disordered and, at best, present local short order with favorable processing conditions. In general, thin films of small molecules are obtained by vacuum evaporation of the materials, whereas polymer thin films are obtained from a deposition from a solution onto substrates by various techniques, such as spin coating, doctor blade, ink-jet printing, roll-to-roll, etc. The former is a long process giving well-controlled and high-quality thin films, while the latter is a rapid process producing less-controlled films of lower quality.

1.1. Organic semiconductors

In general, crystalline solids are semiconductors. Due to the covalent bonds between atoms, which are periodically arranged in the solid, the carriers in these materials have high charge mobility since the electrons and holes are delocalized and can move freely in the allowed energy bands under an applied electric field. The electrical conductivity of semiconductors having a high charge carrier density by doping or by thermal generation is consequently important. Depending on their energy, free electrons occupy states in the conduction band (CB) whose minimum energy level is denoted as E_C, and free holes occupy states in the valence band (VB) whose maximum energy level is denoted as E_V. The *band gap* of the semiconductor is defined as the difference between the two energy levels:

$$E_G = E_C - E_V \tag{1.1}$$

Common polymers are electrical insulators due to their large band gap, which allows very few free charge carriers to be formed at ambient temperature. In polymers containing alternating single and double carbon bonds, called *conjugated polymers*, the adjacent carbon atoms form covalent σ-bonds and their hybrid p-orbitals form π-bonds. The fact that the bond arrangement can alternate between the carbon atoms will make the delocalization of the electrons (π-electrons) possible. The delocalized electrons can be shared by all the other carbon atoms covalently bonded, as shown in Figure 1.1. In other words, they can move along the molecular backbone and within the polymer volume where the molecule distribution is continuous. The extent of the conjugation of the π-electron system is called the *conjugation length* (or the length of the chain of single and double carbon bonds).

According to the model of the molecular orbitals, when a solid is composed of many molecules which interact with one another, the superposition of the orbitals will result in an energy splitting and the formation of energy bands, which are similar to those observed in inorganic semiconductors. The energy of the most energetic π-electrons in the VB corresponds to the *highest occupied molecular orbital* (HOMO) and that of the less energetic π-electrons in the CB corresponds to the *lowest unoccupied molecular orbital* (LUMO). These orbitals are comparable respectively with the E_V and E_C energy levels in inorganic semiconductors, and the energy difference between the LUMO and the HOMO levels corresponds to the energy gap E_G of the organic semiconductor (Figure 1.2).

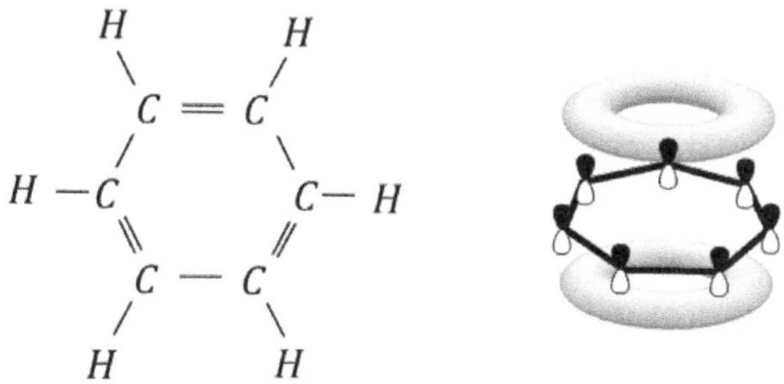

Figure 1.1. *Delocalization of π-electrons in a single carbon ring benzene molecule*

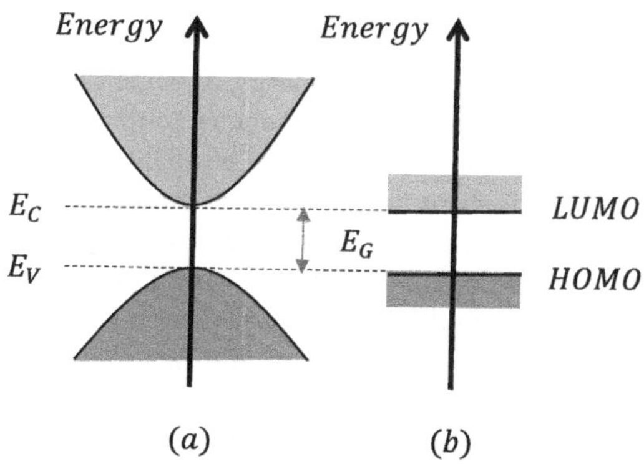

Figure 1.2. *Schematic representation of characteristic energy levels in (a) inorganic semiconductor and (b) organic semiconductor. For a color version of this figure, see www.iste.co.uk/nguyen/defects.zip*

Despite this similarity, the semiconductor character of organic materials is not identical in many aspects to that of conventional semiconductors like silicon.

For the transport process, the HOMO and LUMO bands in organic semiconductors are narrow (<0.5 eV) as compared with those of their inorganic counterparts (<5 eV). A band-like transport of charge carriers is then unlikely, except for very pure organic crystals. Instead, thermally assisted hopping between localized states in the band gap seems to be the most adapted transport mechanism for organic semiconductors. Because of their particular disordered structure, organic materials are prone to generating a high density of defects, which introduces localized states into the energy band gap. These states are distributed and generally described and modeled by a Gaussian function. This state organization is different from the conventional semiconductors, in which defect states are generally located on discrete energy levels. The density and the energy of the Gaussian distribution depend on the semiconductor and the structure of the organic device, using it as an active component. The distribution may evolve due to the effects of the degradation of the semiconductor. The charge transport by the hopping of carriers from molecule to molecule results in a low charge mobility in organic semiconductors and high resistance in organic devices. For comparison, the high limit of mobility in organic semiconductors is estimated to be $\sim 10\ cm^2V^{-1}s^{-1}$, whereas in inorganic counterparts, it is usually higher than 1,000 $cm^2V^{-1}s^{-1}$.

For the optical processes, the chemical bonding between molecules in organic semiconductors leads to high dielectric constants ($\varepsilon_r > 10$) as compared to those of inorganic materials ($\varepsilon_r < 3$). This property impacts on the exciton transport in the solid, and in turn the optical processes in the materials and devices. In short, an exciton or a bound electron–hole pair is created by the absorption of an incident photon. For organic materials, an exciton is a mobile molecular excited state in which the electron is excited to the high-energy S_1 state and the hole is at the S_0 ground state. It can diffuse inside the semiconductor until its dissociation into a free electron and a free hole by diverse processes. The dielectric constant being low in inorganic semiconductors, the excitons have a high binding energy (~500 meV) and are then stable and able to diffuse over a long distance. In contrast, in organic materials, the low constant dielectric leads to a low binding energy (~10 meV) of the exciton, which may be dissociated after diffusion over a short distance (or *diffusion length*). Therefore, if needed, the materials should be structured so as to allow for excitons to be dissociated and charge carriers to be collected over a distance shorter than their diffusion length. An interesting feature of organic materials is the possibility to fine-tune the band gap to the desired magnitude with chemical changes of their structure, for instance by incorporation of functional groups to the molecules to change their LUMO and HOMO levels and then modify the band gap (Cheng et al. 2009). This technique is often used for obtaining the desired emission color in light-emitting diodes and also the band gap of donor–acceptor structures in solar cells.

1.2. Doping of organic semiconductors

Doping a conventional semiconductor consists of introducing impurities into the intrinsic material in order to increase the density of the free charge carriers (electrons or holes), which in turn enhance the electrical conductivity of the sample. The impurities or dopants are materials that contain atoms having one valence electron more (called *donors* for an N-type doping) or one valence electron less (called *acceptors* for a P-type doping) than the atoms of the semiconductor. By providing a small amount of energy to the semiconductor, the donor atoms are negatively ionized and each of them donates an electron to the CB of the material.

Similarly, the acceptor atoms are positively ionized and each of them receives an electron from its neighboring host atoms, which creates a hole in the VB of the material. By controlling the dopant concentration, it is possible to adjust the free charge carrier density in the conduction and the VB for obtaining the desired conductivity.

The principle of doping of organic semiconductors is similar due to the incorporation of electrons or holes to the host material using an *electron donor* (N-type doping) or an *electron acceptor* (P-type doping) material. The choice of the dopant molecules depends on the HOMO and LUMO levels of the host material. For an N-type doping, an electron should be transferred from the VB of the donor molecule to the CB of the host molecule. This is possible only if the *ionization energy* (IE) of the donor is low enough relative to the vacuum level to position its HOMO level closely above the LUMO level of the host molecule. Similarly, for a P-type doping, an electron should be transferred from the VB of the host molecule to the CB of the acceptor molecule. This is possible only if the *electron affinity* (EA) of the acceptor is deep enough to position its LUMO level closely below the HOMO level of the host molecule (Figure 1.3).

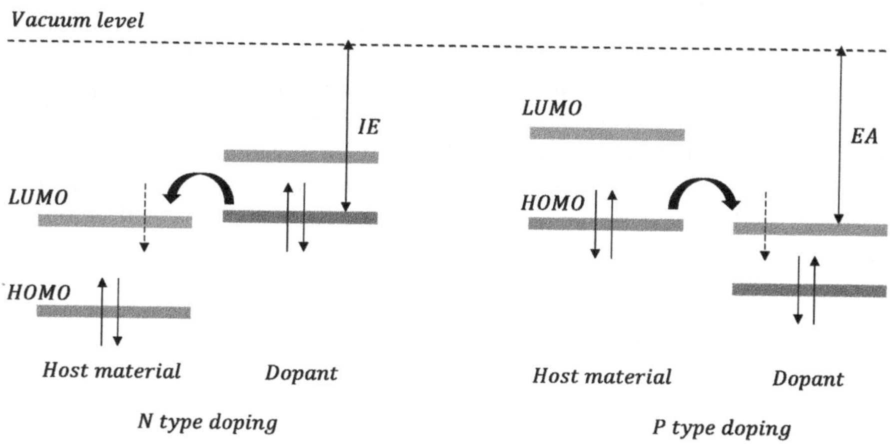

Figure 1.3. *Schematic doping process in organic semiconductor in: a) N-type doping, b) P-type doping. EA is the electron affinity and IE is the ionization energy of the organic material. For a color version of this figure, see www.iste.co.uk/nguyen/defects.zip*

The doping efficiency defined as the ratio of the concentration of free charge carriers $n_{e,h}$ to the concentration of dopants n_d is usually lower than unity. However, due to the high concentration of free charge carriers in the allowed energy bands, doped semiconductors have high conductivity and the charge transport is improved in devices. The doping can be applied to all the organic layers, that is, the active layer and the transport layers, and allows for reducing losses in the injection barriers at the interfaces, namely at the electrode/organic layer contacts. N-type- and P-type-doped semiconductors are currently used in *p–i–n* multilayer structures where *p* is a hole-rich layer and *n* is an electron-rich layer, which sandwiches an intrinsic active layer *i*. It should be noted that doping also produces dopant ions that interact with carriers on the localized states, producing additional deep traps, which reduce the hopping rate of carriers (Arkhipov et al. 2005).

1.3. Organic electronic devices

As some organic materials behave as semiconductors that can be furthermore doped to provide N- and P-type materials, it is straightforward that organic electronic devices can be realized by using the same structure and working principle as their inorganic counterparts. For almost any application, diodes and field-effect transistors are the principal components of the organic devices.

The first studied devices using organic thin films as an active layer were organic light-emitting diodes (OLEDs), which have been used in commercialized applications since 1987. The development of new organic materials, new processing techniques and new device architectures and the understanding of physical processes in OLEDs were then applied to the study of other organic devices. Organic solar cells (OSCs or OPVs), organic field-effect transistors (OFETs), organic thermoelectric devices (OTEs), organic sensors and memories are some of the most intensively studied devices for practical applications in the new field of organic electronics.

1.3.1. *Architectures of organic devices*

1.3.1.1. *Basic architectures*

The basic architecture of the organic diode is that of the *Schottky diode*. It consists of a semiconductor layer sandwiched between two metal electrodes, whose work function is chosen to match with that of the semiconductor in such a manner that the current density is high with an applied forward-bias voltage and negligible with a reverse-bias voltage. Usually, for describing the injection of charge carriers from the electrode to the semiconductor, a *Schottky barrier* is defined as a function of the metal work function Φ_m and the electron affinity (EA) or χ of the semiconductor. For an N-type semiconductor, the Schottky barrier is given by

$$\Phi_{Sch} = \Phi_m - \chi \qquad [1.2]$$

For a P-type organic semiconductor, the work function should be high for the anode and low for the cathode. Early organic devices were built with a structure of metal/organic/metal showing a diode current–voltage characteristic, which is explained by the difference in the nature of the metal/organic contacts.

A more evolutive architecture uses the P–N junction in which the active layer is composed of two layers of N-type and P-type organic semiconductors (Figure 1.4). These structures called *bilayers* have been used for organic solar cells with a low efficiency because of the small interfacial (or junction) area, which limits the concentrations of the dissociated charge carriers (Zhao et al. 2014).

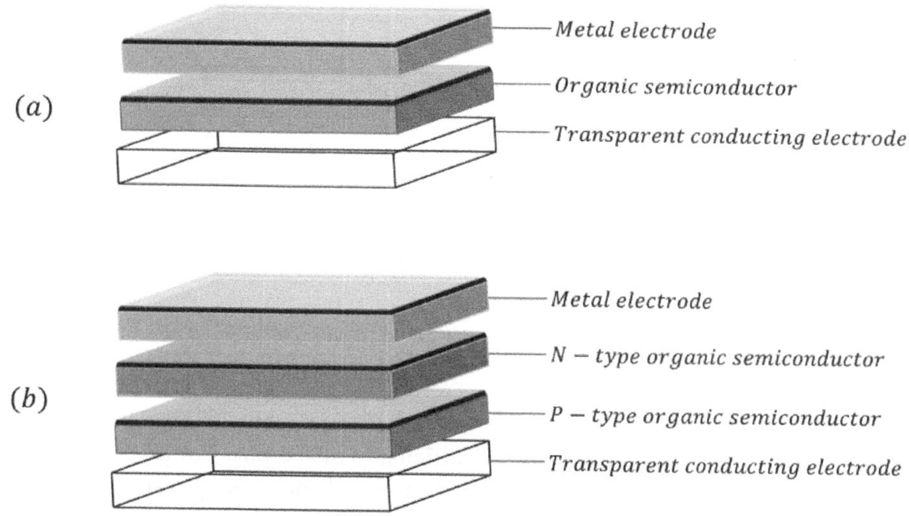

Figure 1.4. *(a) Basic architecture of organic diodes (OLEDs, OPVs) and (b) bilayer structure of organic diodes. For a color version of this figure, see www.iste.co.uk/nguyen/defects.zip*

The basic structure for OFETs is similar to that of inorganic transistors and is composed of an organic semiconductor and three electrodes, which are the source electrode (S), the drain electrode (D) and the gate electrode (G). Depending on the deposition order of the layers, there are four possible configurations of OFETs which are: i) bottom gate, bottom contact (BGBC), ii) bottom gate, top contact (BGTC), iii) top gate, bottom contact (TGBC) and iv) top gate, top contact (BGBC) (Figure 1.5).

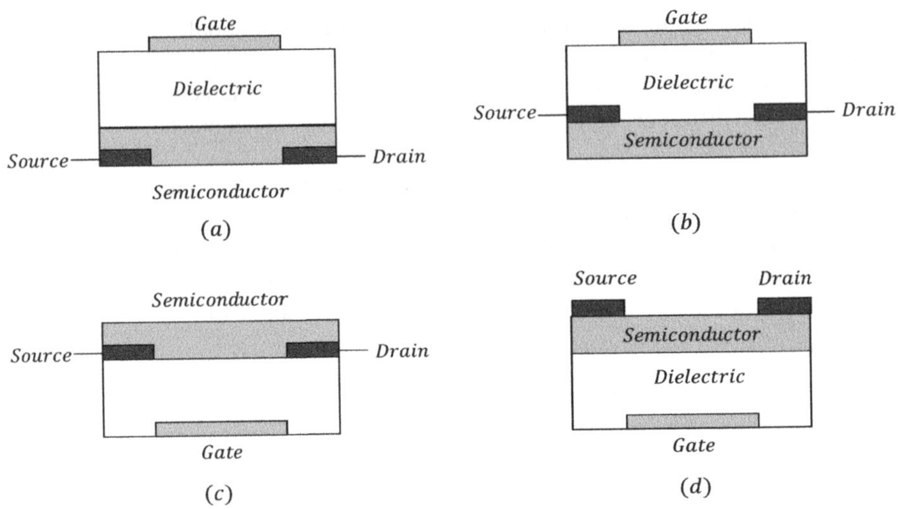

Figure 1.5. *OFET device structures: (a) top gate-bottom contact (TG-BC), (b) top gate-top contact (TG-TC), (c) bottom gate-bottom contact (BG-BC) and (d) bottom gate-top contact (BG-TC). For a color version of this figure, see www.iste.co.uk/nguyen/defects.zip*

1.3.1.2. *Multilayer structures*

Additional layers with different functionalities are often incorporated into the basic structures in order to improve the efficiency of the physical processes in the organic device. Depending on the applications, hence the physical processes occurring in the active layer, the incorporation of these layers is useful or even necessary for some typical devices, but they are not always used in all the devices.

For OLEDs and OPVs, the typical architecture of the devices is composed of three layers sandwiched between two metal electrodes. One of the electrodes is a transparent conducting metal oxide, usually indium tin oxide (ITO) to allow the light to go in or out of the structure. Two additional layers are *electron* and *hole transport layers* (ETL and HTL), which are chosen for the high mobility of their charge carriers and allow for improving the charge transport of carriers between the active layer and the electrodes. The HOMO and LUMO orbitals of the transport layers should be intermediate between the corresponding levels of the electrodes and the active layer in such a manner that the energy barriers between the corresponding layers are small enough to facilitate the injection or extraction of the carriers through low energetic barriers. Furthermore, using doped transport layers leads to an improvement of the conductivity due to the increase of the carrier concentrations and makes the transport more efficient in devices, thus increasing the device efficiency. For illustration, a typical structure of OLEDs is shown in Figure 1.6 with its operation principle.

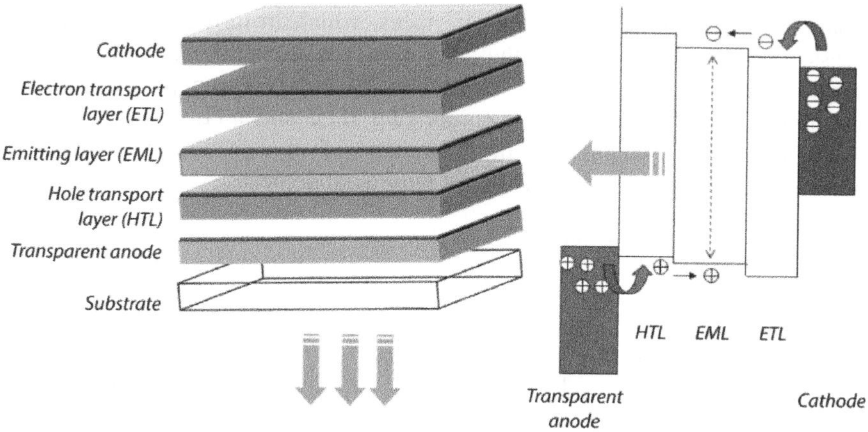

Figure 1.6. *Typical structure of OLEDs with HTL and ETL transport layers and their operation principle. For a color version of this figure, see www.iste.co.uk/nguyen/defects.zip*

Usually, additional layers are used to improve the efficiency of organic devices. In the case of OLEDs, to improve the charge injection from the electrodes to the active or adjacent layer, *electron and hole injection layers* (EIL and HIL) are added at the cathode and anode interfaces. The injection materials are chosen by their HOMO and LUMO orbitals whose energy levels should be intermediate between the corresponding levels of the electrode and the adjacent layer. The injection layers facilitate the charge injection by artificially decomposing the injection barrier into two smaller barriers at the interfaces, which are easier to overcome for the carriers (Figure 1.7).

To improve the charge balance due to the difference in mobilities of charge carriers in the active layer, electron and hole *blocking layers* (EBL and HBL) are added at the HTL and ETL interfaces. The blocking layers are used especially in OLEDs to confine electrons, holes and excitons and limit the escape of carriers from the emission zone and thus improve the recombination in the emitter. The LUMO of the EBL should be located reasonably above the LUMO of the active or *emitting layer* (EML) and its HOMO should be closely aligned with the HOMO of the HTL. Similarly, the HOMO of the HBL should be located reasonably below the HOMO of the EML and its LUMO should be closely aligned with the LUMO of the ETL (Figure 1.7).

The use of multilayers for OLEDs and OPVs allows for improving their charge transport and, consequently, their efficiency. However, it is important to keep in mind that the presence of the additional layers in devices may also drive the formation of various defects at the interfaces, which in turn affects the transport process. Therefore, the choice of the layer materials should be carefully considered in order to fully take advantage of all that these layers can provide. This also explains the widely used structure comprising HTL and ETL transport layers for OLEDs and OPVs, which have the advantage of improved conductivity with a minimized possibility of the formation of interfacial defects.

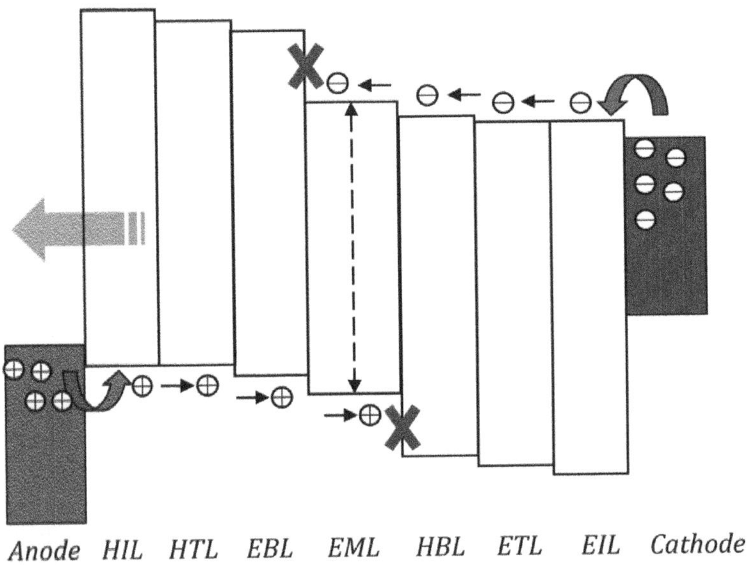

Figure 1.7. *Schematic structure of a multilayer OLED with injection, blocking and transport layers. For a color version of this figure, see www.iste.co.uk/nguyen/defects.zip*

1.3.1.3. *Heterojunction structure*

In bilayer devices, the two semiconductors in contact are the donor (D) and the acceptor (A), which allows for transferring the charge carriers between the materials (*charge transfer*). The efficiency of the process is usually low because, on the one hand, the contact area between semiconductors is limited by the surface of the devices (which limits the recombination in the case of OLEDs) and, on the other hand, because the diffusion length of excitons is generally short ~10 nm (which limits the dissociation and collection of the charge carriers in the case of OPVs). A *bulk heterojunction* (BHJ) structure is obtained by mixing the donor and the acceptor for obtaining a blend, in which the contact area between the semiconductors is dispersed in the volume of the active layer, greatly enhancing the possibility of charge transfer.

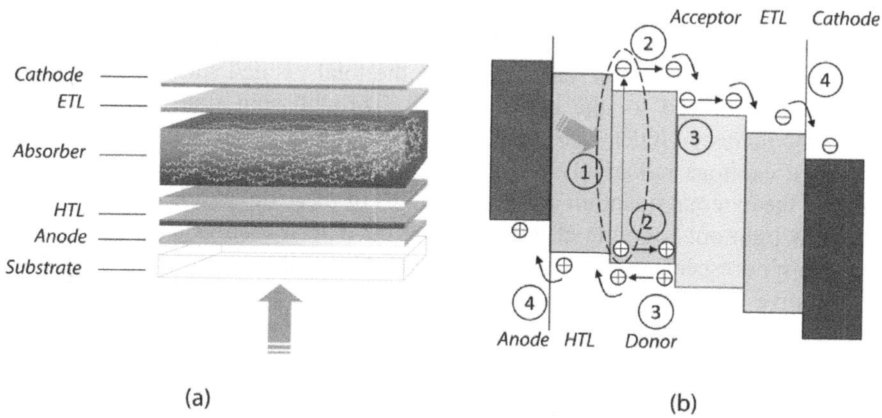

Figure 1.8. *(a) Schematic structure of the BHJ solar cell and (b) photovoltaic effect in a complete solar cell. The different steps of the effect are: 1. exciton generation, 2. exciton diffusion, 3. exciton dissociation and 4. charge carrier transport and collection. For a color version of this figure, see www.iste.co.uk/nguyen/defects.zip*

The BHJ structure is commonly used for preparing the absorber layer of OPVs and allows for significantly increasing the conversion efficiency of these devices (Figure 1.8). However, the film deposition of the blends requires a rigorous processing protocol and a strict control of the parameters for obtaining a good-quality film morphology and uniform distribution of the blend in the bulk of the film, which are necessary for improving the cell efficiency. It is often observed that during thermal stress, the BHJ film morphology becomes unstable and shows large variations in composition and orientation, which in turn affect the transport and the efficiency. Besides these difficulties, there is also a problem of occurrence of phase separation of the blends with time, which introduces additional defects, decreasing the cell performance. The use of scalable nanostructured electrodes such as nanorod arrays (Gonzalez-Valls and Lira-Cantu 2009) can help solve the blend distribution problem.

1.3.2. *Organic light-emitting diodes (OLEDs)*

The working principle of OLEDs is similar to that of inorganic LEDs. Basically, under an applied voltage, the charge carriers are injected from the electrodes, holes from the anode and electrons from the cathode to the emitting layer. Excitons or electron–hole pairs are formed and transported by the electric field to the recombination centers where they radiatively recombine and a photon is emitted to the outer medium through the transparent anode, usually indium tin oxide or ITO (Figure 1.6).

The first generation of OLEDs is based on fluorescent material emitters in which only the singlet states, representing 25% of the total excited states by the injected charges, are involved for the light emission. Through spin statistics, the *internal quantum efficiency* (IQE), or the ratio of the number of photons emitted to the number of excitons radiatively recombined in the emitter, is limited to this upper limit and the external quantum efficiency (EQE) of these devices is at best ~5%, due to the low light out-coupling efficiency of ~20%. The second generation of OLEDs used phosphorescent dyes with fluorescent materials to exploit the normally non-radiative triplet excitons and thus improve the overall device efficiency (Baldo et al. 1998). This is made possible by the use of heavy-metal complexes as dopants, which allows transitions from the triplet excited state to the ground singlet state by spin–orbital coupling. Internal quantum efficiency of nearly unity could be obtained in such devices (Adachi et al. 2001). The efficiency of OLEDs is further enhanced by using emitter materials having *thermally activated delayed fluorescence* (TADF) properties.

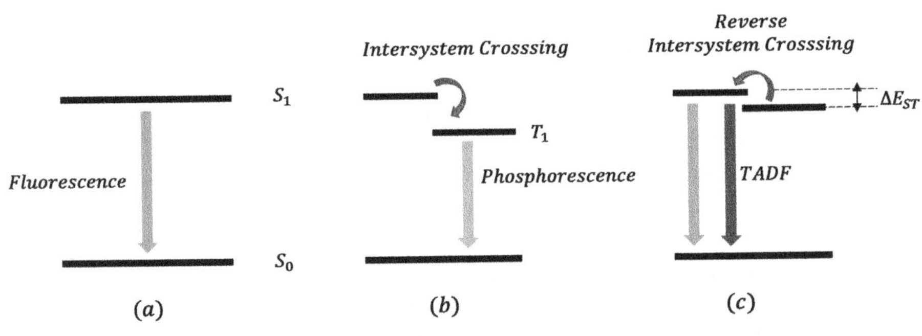

Figure 1.9. *Schematic representation of emission mechanisms in organic emitters: (a) fluorescence, (b) phosphorescence and (c) TADF. For a color version of this figure, see www.iste.co.uk/nguyen/defects.zip*

The energy gap between the singlet and triplet states ΔE_{ST} of these materials is small enough ($<100\,meV$) to allow a *reverse intersystem crossing* (*RISC*) mechanism $T_1 \rightarrow S_1$ between the triplet and the singlet states. This mechanism is a thermally assisted up-conversion process, bringing an electron from the triplet state to the singlet of higher energy and allows it to decay radiatively yielding a fluorescence emission (Figure 1.9). Because the RISC process is slow, the TADF emission is delayed and takes place after the typical emission of singlet excitons created on the S_1 states. The EQE of OLEDs using a TADF emitter can reach 37% for a conventional planar structure (Lin et al. 2016).

OLEDs are principally used for the fabrication of displays and lighting devices. The required luminance for OLED-based displays is >100 cd.m^{-1} which can be realized for screens of small portable devices (cellphones, computers) or of large surface televisions (standard or rollable TV screens). For lighting applications, the required luminance is >1,000 cd.m^{-1} which can be realized for large plane or curved emitting surfaces (luminaires). On the other hand, for an agreeable vision, it is convenient to use white organic light-emitting diodes or WOLEDs. There is no single-molecule emitter that can provide a sufficiently large spectral range to produce a white emission because the white light requires three primary color components: red, green and blue (RGB). WOLEDs therefore have a multilayer structure (stacked or pixelated) using either fluorescent, phosphorescent or TADF emitters. In general, a blue fluorescent emitter is used with a green and a red phosphorescent emitter for preparing a high-efficiency WOLED. Other architectures have also been used for producing white light diodes for instance, blue OLED with down-conversion layers, multiple-doped emission layers, etc.

1.3.3. *Organic solar cells (OSCs or OPVs)*

OPVs convert solar light into electricity following a four-step process (Figure 1.8) described as follows:

– Step 1: in a basic solar cell structure, under appropriate light irradiation. electron–hole pairs or excitons are generated in the active layer or absorber with a significant binding energy (of few tens of *meV*) due to the high dielectric constant of the organic semiconductor.

– Step 2: the excitons diffuse in the absorber over a diffusion length (in general, of ~5–30 nm). The carrier transport takes place by hopping between localized states or by an energy transfer mechanism.

– Step 3: the excitons are dissociated into a free electron and a free hole. Because of the high binding energy in organic materials, excitons cannot be dissociated by thermal agitation but by charge transfer at the donor/acceptor interface.

– Step 4: after dissociation of an exciton, the free electron and hole are transported to the respective electrodes, creating a voltage which is a function of the collected charges.

The EQE for OPVs is defined as the ratio between the collected charges and the number of incident photons. For obtaining a high EQE from an OPV, it is necessary to have a highly absorbing active material, a high diffusion length of excitons, a good separation of the exciton charge carriers and a high ability of charge collection at the electrodes. The early organic solar cells were fabricated using a planar bilayer structure, obtaining a *power conversion efficiency* (PCE) of ~1% (Tang 1986). The PCE is defined as the ratio between the maximum electric power output P_m and the incident light power input P_{in}.

Later, improvements of the cell performance were achieved by introducing the BHJ concept and by using C_{60} fullerene and its derivatives to replace the N-type organic semiconductor. For instance, poly(3-hexylthiophene) (P3HT)-[6,6]-phenyl C61-butyric acid methyl ester (PCBM) blends have been intensively studied for improving the efficiency of BHJ solar cells. Using a P3HT:PCBM blend as the absorber, a PCE of ~ 5.4% has been realized (Laird et al. 2007). Development of new donor polymers made it possible to improve the performance of BHJ based on fullerene derivative acceptors and to obtain efficiencies higher than 11% (Zhao et al. 2016). From the investigations of BHJ devices, it is also found that the morphology of the blend, and in particular the donor–acceptor interface, is of primary importance for the achievement of high device efficiency. It is often observed that the morphology of the blend using fullerene acceptors is sensitive to the film processing conditions and environmental stresses, undergoing changes during the device operation and enabling the formation of polymer and fullerene-rich regions. This phase separation results in a drop in the cell efficiency.

For replacing fullerene derivative acceptors, a new class of organic materials, called *non-fullerene acceptors* or *NFA*, has been introduced and developed as acceptors, which are associated with donor polymers to form the photoactive layers. They offer two main advantages over fullerene derivatives. On the one hand, by modifying the chemical structure of the material by molecular design, it is possible to finely adjust the energy levels and the band gap for adapting or improving the light absorption of the absorber. On the other hand, using NFA provides blends of high stability as compared to the fullerene-based acceptors by mitigating the phase separation process. Additionally, the NFA blends are more economical to produce than fullerene ones and can potentially reduce the cell fabrication costs. Single-junction OPV using a small-molecule acceptor ITIC associated with a polymer donor PBDB-T as an absorber blend shows a PCE of 13.1% (Zhao et al. 2017). By using an acceptor Y6 blended with a polymer donor PBDT-TF and used for the absorber of OPVs, the solar cells show a PCE higher than 18% (Cui et al. 2020). Improvement of the efficiency and the stability of the devices can be achieved with the use of *ternary blend bulk heterojunction* where two donors and one acceptor ($A: D_1: D_2$) or two acceptors and one donor ($D: A_1: A_2$) are blended to form the absorber (Yang et al. 2013). The choice of the third element can allow for broadening and enhancing the light absorption, thus increasing the device efficiency. Moreover, its incorporation contributes to optimize the structure and the morphology of the blend and improves the stability of the cells. Single-junction using a ternary blend of a polymer donor PBQx and two acceptors BTP-eC9-2Cl and a BTA3 derivative shows a PCE of 19% (Cui et al. 2021).

Beside the organic semiconductors used for the absorber layer of a solar cell, the hybrid organic–inorganic perovskite is the most promising materials for emerging photovoltaic technologies. The chemical formula of perovskites is ABX_3, where A (generally methylammonium $CH_3NH_3^+$) and B (Pb^{2+}, Sn^{2+}) are cations occupying different sites, and X is an anion (I^-, Br^-, Cl^-). The perovskites have several remarkable physical properties that make them highly efficient when used as an absorber. The key features of the materials include a strong light absorption, a low non-radiative recombination, a high carrier mobility and a long carrier diffusion length. In addition, thin films of perovskite can be deposited by the solution process, which would enable the mass and low-cost production of solar cells. The efficiency of the very early perovskite-based solar cells was ~2% using $CH_3NH_3PbBr_3$ as an absorber (Kojima et al. 2006). Development and improvement of the perovskite and the transport materials and optimization of the properties of the interface between the active layer and the electrodes have allowed us to progressively and significantly increase the cell efficiency, which has reached 25.5% (Min et al. 2021).

Both recent organic and perovskite solar cells have high conversion efficiency that is comparable to that of inorganic counterparts. The negative aspect for their use is the lack of high stability that needs to be strongly improved for future commercialization.

1.3.4. *Organic field-effect transistors (OFETs)*

The operating mode of OFETs is similar to that of inorganic MOSFETs, which is based on the use of the voltage V_{GS} applied between the gate G and the source S to create an electric field in the dielectric in order to modulate the conductivity of the channel between the source and the drain D. The main difference between OFETs and MOSFETs is that the organic semiconductors used are intrinsic, while the inorganic ones are doped. This implies that OFETs are based on the principle of *charge carrier accumulation* in contrast to MOSFETs that are generally based on the principle of *charge carrier inversion*. Upon the application of the gate voltage, a conducting channel is formed at the semiconductor/dielectric interface due to the accumulation of charge carriers and a current I_{DS} will flow between the source and the drain electrodes when a voltage V_{DS} is applied. The output characteristic $I_{DS}(V_{DS})$ (Figure 1.10) shows two working regimes. In the linear regime, the drain current increases linearly with the drain–source voltage. At a critical point where $V_{DS} = V_{GS} - V_T$, the channel becomes pinched off and the I_{DS} is saturated since any further increase in V_{DS} will result in growing the depletion region near the drain electrode.

The drain current depends on the carrier density and their mobility. Therefore, the performance of the transistor will be strongly dependent on the carrier mobility, which is the key characteristic of the organic semiconductor.

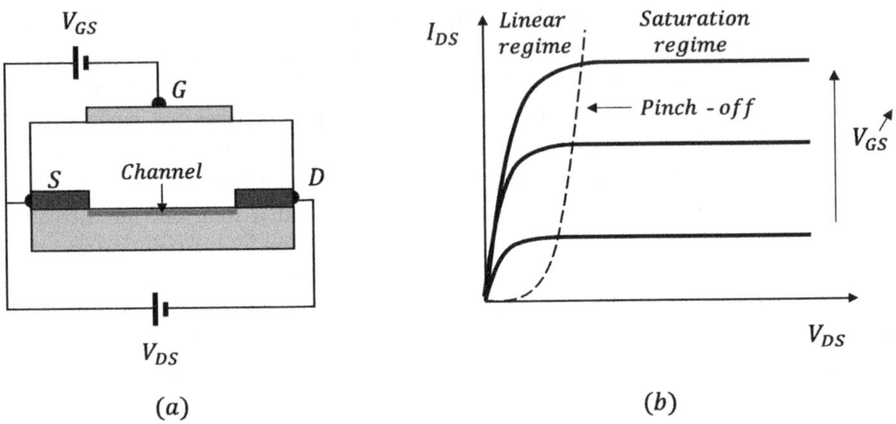

Figure 1.10. *(a) Schematic representation of the operating mode of OFETs. (b) OFET output characteristic. For a color version of this figure, see www.iste.co.uk/nguyen/defects.zip*

In the first OFET devices using polythiophene as the semiconductor, the measured mobility was 10^{-5} cm^{-2}V^{-1}s^{-1} (Tsumura et al. 1986). Progressively, improvement in material quality and processing has led to higher carrier mobility exceeding 10 cm^{-2}V^{-1}s^{-1}. By increasing the degree of order of the organic semiconductor (PCDA polymerized crystals) used in OFETs, the field-effect mobility of carriers up to 50 cm^{-2}V^{-1}s^{-1} has been observed (Yao et al. 2017). These mobility values would fulfill the requirements for voltage-driven applications such as OLED active matrix addressing.

Most of the organic semiconductors used in OFETs are of P-type, forming P-channel accumulation mode transistors. A few organic materials such as C60 fullerene can, however, form N-channel transistors. Employing both P-type and N-type materials allows for achieving N and P channel operations in a single transistor. Such a transistor is called an *ambipolar OFET* and can operate as an N-type or P-type transistor depending on the polarity of the gate voltage V_{GS} (Dodabalapur et al. 1996). The presence of both electrons and holes in the active layer of ambipolar OFETs makes it possible to achieve light emission in the devices like a light-emitting diode. Under appropriate bias conditions, the transistor works as an LED and also as a switch and is called an *organic light-emitting transistor* (OLET) (Hepp et al. 2003a). The active layer is the transport and the emission layer in which light emission occurs in the conducting channel and no transparent conducting electrode is needed.

Often, operational OFETs show low output current and high working voltage (in the order of μA and a few dozen volts respectively) because of the low conductivity of the organic semiconductor, and this is a disadvantage for practical applications. In order to improve these characteristics, a vertical structure of the organic transistor defined as a VOFET is realized (Ma and Yang 2004). The structure of a VOFET comprises a capacitor and an OFET with a common source electrode of a rough surface, separating the active layer and the dielectric (Figure 1.11).

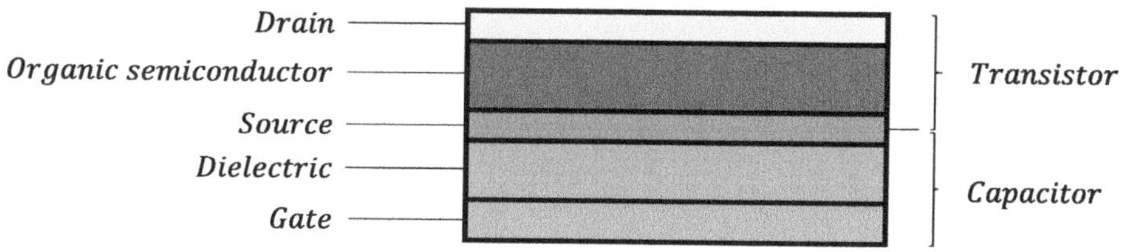

Figure 1.11. *Schematic representation of the structure of a vertical organic field-effect transistor (VOFET). For a color version of this figure, see www.iste.co.uk/nguyen/defects.zip*

The capacitor of high capacitance facilitates the charge carrier injection from the source to the semiconductor layer when appropriate gate voltage is applied to the device and increases the drain current. Furthermore, with this structure, the channel length is short, and the cross-sectional area is large, the current density is then high (~4 $A.cm^{-2}$) and the operating voltage is low (< 5 V), producing a high switching speed of the transistor. New architectures of VOFETs have thus been developed, providing very high current density (~3×10^{6} $A.cm^{-2}$) (Lenz et al. 2019) that can be exploited for voltage driving of OLED displays.

In this chapter, a short review of the electronic structure of organic semiconductors is presented, and the different architectures of the most popular organic devices including OLEDs, OFETs and OPVs are described. These notions will be used to investigate the defect processes in organic materials and devices in the next chapters in order to understand their formation mechanisms and their effects on the operation of devices. As there are a large number of organic materials that are investigated as active components in devices, it would be difficult to analyze defects in all these materials and compare the results obtained from devices made with semiconductors of different structures. Instead, studying the defects in several typical semiconductors in use in specific devices seems to be more consistent to understand the defect–property relation. Consequently, for most of the examples given in the following chapters, the results obtained in OLEDs, OFETs and OPVs will be given for known conjugated polymers (PPV and derivatives) and small molecules (Alq3), as well as donor/acceptor blends (P3HT:PCBM) and hybrid perovskites that are used in these devices.

2

Defects in Materials

A defect can be defined as a fault in a material or a device that causes it not to work correctly. In other words, if defects are present in the material or device, their original properties will be affected or changed, which in turn will impact the functioning or the operation for which they are designed. Ideally, a material should be perfect: that is to say, containing no defects. But real materials always have defects, whatever the care provided during synthesis and processing. In some cases, the defect formation can be identified but more generally, the formation process still remains a complex subject, needing extensive studies to be fully understood.

In order to investigate the defects in organic semiconductors, we will first study the defects in known inorganic semiconductors: crystalline materials. Then, by considering the defects in amorphous semiconductors, whose structure is close to that of organic materials, we will examine the possible formation processes of the latter and relate their properties to those studied in both crystalline and amorphous semiconductors.

2.1. Order and disorder

The concept of order–disorder has been considered in numerous fields of physics. For materials, the term *ordered* is used to describe material lattice sites grouped into sublattices, each of which is occupied by one kind of atom or various groups of atoms, whereas in *disordered* materials, the grouping is not possible since each sublattice is occupied by the constituents at random. As the ordered structure may not be infinite in materials, the system is said to have *long-range order* whenever there are correlations or symmetries between an atom on a given site and another atom on a finite distant site. In a crystal, if a one-unit cell is known at a lattice point, then all the atomic positions at any other lattice point can be determined through translational symmetry. On the other hand, in disordered materials, when atoms on sites close to each other still have correlations or symmetries, the system is said to have *short-range order*. The atomic arrangement can only be determined over a short distance of the order of one or two atomic spacings. The concept of long-range order and short-range order is very often used when studying the physical properties of organic semiconductors, as we will see in the next section.

2.2. Crystalline semiconductors

Typical inorganic semiconductors such as silicon (Si) or germanium (Ge) are crystals, that is, materials having atoms (or molecules) built up by periodic arrangement of atomic arrays in the three-dimensional space. Ideal crystals are defined as materials having long-range order and infinite translational symmetry. The energy bands of such semiconductors are well defined with valence and conduction band edges E_V and E_C and hence a well-defined band gap E_G between these two energy levels. The *density of states* (DOS) describing the distribution of energy states within the bands is proportional to $E^{1/2}$, where E is the energy of an electronic state (Figure 2.1). Any energy state located inside the DOS can be occupied by a charge carrier and is termed as a *delocalized* or *extended* state.

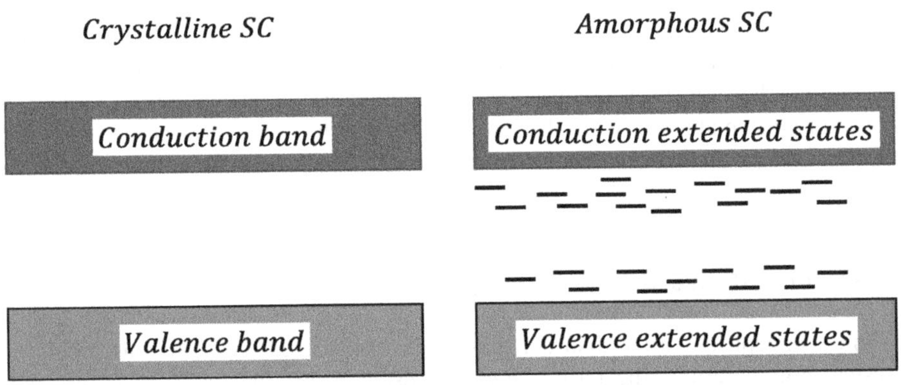

Figure 2.1. *Schematic representation of the energy band diagram of crystalline and amorphous semiconductors. For a color version of this figure, see www.iste.co.uk/nguyen/defects.zip*

In real semiconductors, the lattice is not perfect and contains imperfections or defects which will interrupt the crystal periodicity and hence the potential in which the charge carriers are transported. Therefore, defects affect the electrical properties of semiconductors.

There are three basic types of defects in crystalline semiconductors (Figure 2.2):

– *point defects* including vacancy (missing atom from a particular lattice site) and interstitial (atom located between lattice sites);

– *chemical impurities* including substitutional (impurity located at the normal lattice site) and interstitial (impurity located between lattice sites) atoms;

– *extended defects* like line dislocations, stacking faults, which extend to several lattice sites and rows of atoms.

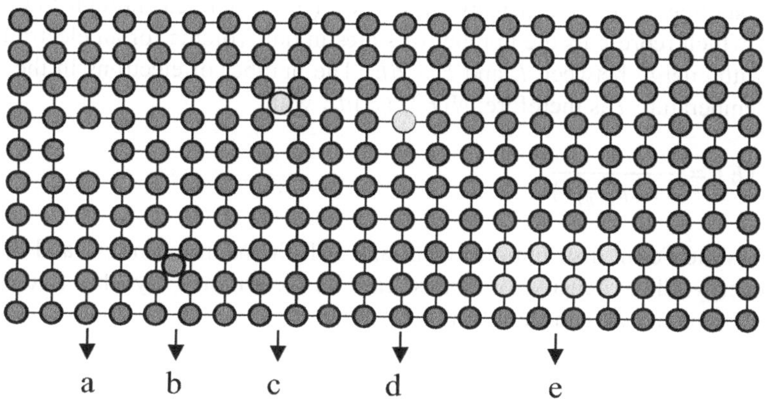

Figure 2.2. *Schematic representation defects in crystalline semiconductors: a) vacancy, b) interstitial, c) interstitial impurity, d) substitutional impurity and e) substitution impurity cluster. For a color version of this figure, see www.iste.co.uk/nguyen/defects.zip*

2.2.1. *Localized states*

In a perfect crystal, all wave functions $\Psi(r)$ are extended, that is, they have the same magnitude at any lattice point in the crystal. In addition, the Ψ have phase coherence thanks to the long-range order. Due to the perfect translational order in the crystal, electron energy is organized in continuous allowed bands with sharp and well-defined band edges.

When an imperfection is introduced into the lattice, the wave functions are affected by the local change in potential at the vicinity of the defect. If the potential change is weak, the wave function remains practically unchanged and the electron energy is not impacted, keeping the energy bands intact. On the contrary, if the potential change is higher than a critical value, a *localized* or *bound state* is created in the forbidden band. This state results from a split of an energy state of the allowed band, under the repulsive or attractive potential introduced by the defect. Each type of defects creates its localized states and their energy level in the band gap.

Localized states are not present in the structure of ideal periodic crystals but only in those containing imperfections or those affected by disorder.

2.2.2. *Density of states (DOS)*

In crystalline solids, the density of states $N(E)$ is defined through the product $N(E)dE$ which represents the number of states in unit volume for an electron in the system with energy between E and $E + dE$. The number of electrons in the energy dE at a temperature T is therefore: $N(E)f(E)dE$ with:

$$f(E) = \frac{1}{1+exp\left(\frac{E-E_F}{kT}\right)} \quad [2.1]$$

$f(E)$ is the Fermi distribution function, k is Boltzmann's constant and E_F is the Fermi energy. The density of states can be determined by the free-electron approximation, and we can show that:

$$E = \frac{\hbar^2 k^2}{2m} \quad [2.2]$$

where $\boldsymbol{k}$ is the wave vector and m is the effective mass. It can be shown then that the electron density of states is given by:

$$N(E) = \frac{(2m)^{3/2} E^{1/2}}{2\pi^2 \hbar^3} \quad [2.3]$$

The band energy diagram of a perfect crystal is given in Figure 2.1 where the band edges are well defined by the energy levels E_C (bottom of the conduction band) and E_V (top of the valence band). The DOS function of a perfect crystal describing the distribution of electronic states in the allowed band is schematically shown in Figure 2.2(a). In real crystals, defects are present and introduce localized levels in the band gap. These levels are discrete if the defects are of the same types.

2.3. Amorphous semiconductors

Amorphous materials differ from crystalline semiconductors by the absence of long-range order in their atomic structure. They are also called non-crystalline or disordered solids because they have no periodic or translational symmetry.

Indeed, amorphous semiconductors can be considered as two specific forms of crystalline semiconductors: a disordered (e.g. amorphous silicon a-Si) and a glassy (e.g. amorphous selenium a-Se) form. In both forms, long-range order is absent but short-range order is preserved. In this configuration, the bonding between atoms is covalent in glassy semiconductors and atoms in disordered semiconductors are tetrahedrally coordinated, which is similar to the network structure of crystalline semiconductors such as silicon. It should be highlighted that the coordination of atoms is generally different from that of the corresponding crystalline material. In addition, bond lengths and bond angles are not constant. Their variations contribute to the overall disorder of the amorphous materials, which contain an important number of defects (intrinsic defects such as broken bonds, unsaturated bonds, voids, foreign atoms and extrinsic defects such as chemical impurities, contaminants) and consequently, their density of states is different from that of crystalline semiconductors.

2.3.1. *Localized states*

Besides defects such as vacancies or interstitials, most defects found in amorphous semiconductors are due to the presence of dangling bonds. A *dangling bond* is defined as an unsatisfied valence on an immobilized atom. In a solid, atoms are organized to develop covalent bonds with their neighbors to attain a stable configuration, that is, with a filled valence shell. For main-group elements, the octet rule applies and states that each stable atom has eight electrons in its valence shell.

As atoms are randomly distributed in an amorphous structure, many atoms have no sufficient bonding neighbors and develop dangling bonds. They are referred to as *immobilized free radicals*. For instance, in a-Si, each immobilized free radical has only three nearest neighbors, and has an unpaired electron in a localized orbital. A dangling bond is electrically neutral (state D^0) but can become charged upon doping (states D^+or D^-). To complete its bonds, the radical tends to capture available electrons or give up its unpaired electron, and acts either as an acceptor or as a donor. A dangling bond state can be considered as equivalent to a state of an isolated atom as it is not bonded. Because of the charge neutrality condition, the electronic states of dangling bonds are localized close to the middle of the band gap. Furthermore, there will be few or no orbital overlap between these states because the dangling bond sites are sufficiently distant from each other. Consequently, the electronic conduction in solids with a high density of dangling bonds can occur through hopping or tunneling between the localized states.

Other localized states due to the disordered structure of the amorphous materials exist and are explained below.

2.3.2. *Density of states (DOS)*

For amorphous materials, the energy distribution of electronic levels is not affected by the absence of long-range order (Davis 1985). In other words, the allowed energy bands including valence and conduction bands separated by a band gap exist like in crystalline semiconductors. On the contrary, because of the disorder of the structure, the DOS of amorphous materials is subsequently modified. To describe long-range order in the crystalline structure, the periodic potential is modeled by an array of square-well potential (Kronig–Penney model) and analytic solutions to the Schrödinger equation make it possible to establish the band energy diagram of crystals. The disorder is introduced through the random variation of the potential wells defined by two parameters V_0 and B (Anderson model: Anderson 1958; Mott 1980): V_0 is the extent of the random potential and B is the bandwidth when there is no potential variation ($V_0 = 0$). From this model, which can apply to heavily doped and amorphous semiconductors, when the fluctuating potential becomes significant: $BV_0^{-1} < 1$, the mean free path λ decreases and when it is lower than the distance of the wells, the states within the band become localized. On the contrary, if the potential variation is small, $BV_0^{-1} > 1$, all states will be extended and there will be only a small change of the band edges: the sharp edges of the DOS are replaced by an exponential tail. In weakly disordered systems, both extended and localized states are present in the band diagram of the semiconductors (see Figure 2.1(b)).

The proposed band model is later modified by Cohen et al. (1969) who considered that the band tailing extends largely into the band gap of the semiconductor and that the overlap of the conduction and valence bands occurs near the middle of the gap, in such a way that the Fermi level is pinned at the overlapping point (Figure 2.3). The electron mobility decreases for energy levels lower than a certain energy E_C^m, and the hole mobility decreases for energy levels higher than a certain energy E_V^m. The transition between these states is a critical energy level called *the mobility edge*, at which localization occurs in the band tail. There are two mobility edges at the bottom of the conduction band E_C^m (electron mobility edge) and at the top of the valence band E_V^m (hole mobility edge), which defines the *mobility gap* E_G^m.

In the scheme of the DOS as a function of energy, the optical gap of the amorphous semiconductor is defined by the extrapolation of the delocalized states E_C and E_V.

$$E_C^{opt} = E_C - E_V$$

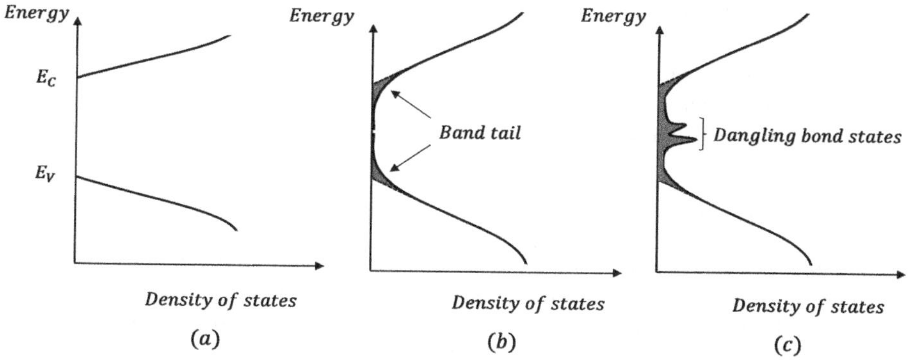

Figure 2.3. *Schematic representation of the density of states (DOS) in (a) a crystalline semiconductor, (b) a heavily doped semiconductor doped and (c) an amorphous semiconductor*

2.4. Organic semiconductors

When studying the transport process, the disorder in organic semiconductors is classified into two categories: diagonal and off-diagonal disorder.

Diagonal disorder (or *energetic disorder*) relates to fluctuations of the energy of molecules or chain segments of the materials, while *off-diagonal disorder* (or *spatial or configurational disorder*) relates to fluctuations of interactions between molecules or chain segments due to their orientation and relative position from each other.

In organic materials, due to their mechanical properties, fluctuations in the torsion angles of chains or molecules are numerous. In addition, the chain lengths in polymers are not constant and are usually distributed via a Gaussian function. Both types of fluctuations contribute to the overall diagonal disorder, that is, the variation of the energy of the HOMO and LUMO levels of the molecules and segments. The energy fluctuations are also affected by the potential variation of the molecules or segments, introduced by impurities in their vicinity. The diagonal disorder in organic materials is similar to the disorder described by the Anderson model in amorphous semiconductors.

Off-diagonal disorder depends on the organization of molecules or chain segments in the materials, which influence their interactions, and thus the electronic couplings in the materials. By controlling the deposition conditions, the quality of the organic material can be modified and improved, which thus mitigates the effects of disorder.

2.4.1. *Polymer structure*

The molecules in polymers can be organized into a network. Indeed, the different chains of a polymer can be linked together by introducing chemical linkages (cross-links) between monomers. In Figure 2.4, the different elements of a polymer network are shown:

– *Strand*: a portion of the polymer chain that links two different chains.

– *Cross-link* (or junction): a group of at least three strands having the same origin on a polymer chain. The functionality of a junction is defined as the number of connected strands (trivalent cross-link for three connected brands or tetravalent cross-link for four connected brands).

– *Loop*: a section of the chain that begins and ends at the same cross-link on a polymer chain.

– *Dangling chain*: a polymer chain that starts at a junction located on another polymer chain and ends by a *chain end*, without being connected to any other chains. A dangling chain is free to relax. Each polymer chain has two chain ends, and high-molecular-weight polymers contain more chain ends than low-molecular-weight polymers.

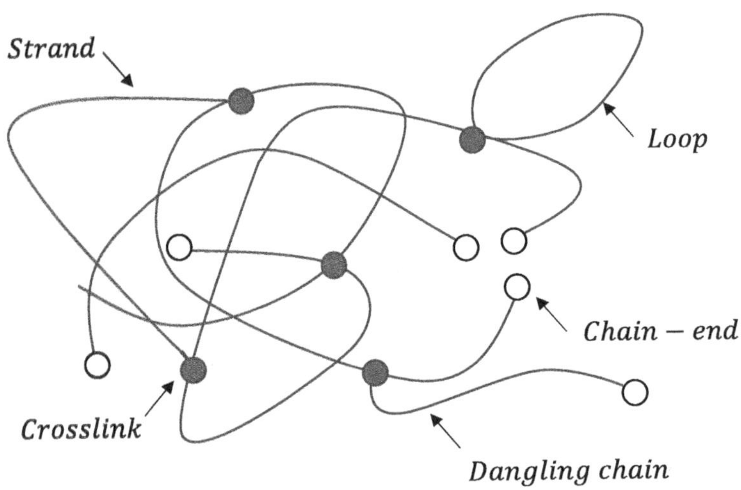

Figure 2.4. *Schematic representation of the polymer network elements*

Each polymer chain is composed of a number of segments separated by twists or kinks. Each segment constitutes an independent entity, and the distribution of the segments will define the electronic state distribution. In a polymer, not all chains are perfectly aligned with each other over the whole size of the solid. Instead, these chains can be considered as aligned in small regions or *crystallites* of average dimensions which are comparable to their conjugation length (few tens of nanometers). The polymer solid can then be described as an ensemble of crystallite regions connected between them by amorphous regions, formed by coiled or tangled polymer chains. The energetic state of electrons in segments is affected by the segment distribution (through the polarizability of the surrounding environment) and the chain distribution (through the interactions with the neighboring chains).

2.4.2. *Polymer crystallinity*

Many polymers show partial crystallinity, proven by their X-ray patterns, which consist of sharp and well-defined features of crystalline regions and smooth varying intensity features of amorphous regions. Both ordered and disordered phases exist in these polymers. A semicrystalline material consists of two different parts: a crystalline part having an ideal crystal structure and an amorphous part having the ideal melt structure. The crystallinity parameter used to characterize the degree of order or disorder is defined as the *degree of crystallinity* whenever the distinction between the limiting states can be clearly established. Several techniques such as wide-angle X-ray scattering (WAXS) have been used to determine the degree of crystallinity of semicrystalline polymer with more or less success (Young and Lovell 2011). However, this concept can hardly apply for real polymer materials because of the presence of defects in their structure, which modify the limits of ordered and disordered portions in the materials.

To better define the intermediate state between crystalline and amorphous materials, the concept of *paracrystallinity* is proposed (Hindeleh and Hoseman 1988). The starting point of the concept is the statistical variation of the dimensions of the unit cell, which is contrary to what happens in a crystal, where the unit cell has the same distance statistics with respect to its nearest neighbors. The consequence of the variation is the loss of the long-range order because of the overlapping of the adjacent lattice sites as the distance increases. In this situation, the deviation of the unit cell depends on the position of neighboring units and does not result from the displacement from its ideal or predetermined lattice site. Consequently, the atom occupying a lattice site will not have equivalent properties as those of its neighboring atoms, and a displacement disorder or defect is created in the material.

To characterize this type of disorder, a parameter called the *paracrystalline distorsion g value* is defined to describe the variation. It expresses the deviation of the unit cell vector in a given direction and describes the variations Δ of the paracrystalline distortions relative to the average distance a between two adjacent structure sites. It can be written as:

$$g = \frac{\Delta}{a} \qquad [2.4]$$

The g value can be experimentally determined from the X-ray diffraction line width obtained from the material analysis (Wunderlich 1973) and makes it possible to estimate the crystalline and the amorphous portions of the semiconductor. Materials with a g value below 1% are crystalline and with a g value between 1% and 10% are paracrystalline (e.g. polymers). Glasses have a highly disordered structure and high g values between 10% and 15% (e.g. SiO_2).

The disorder in bonding leads to a change of the local charge distribution and consequently creates fluctuations in the potential, which affects the localization of electrons.

2.4.3. *Defects in conjugated polymers*

Conjugated polymers can be considered as an assembly of ordered and disordered regions with no well-defined distribution in general. Structurally, they are mixtures of molecules that differ in size and distribution of repeated units, from which disorder is introduced. Consequently, they possibly contain both types of defects that can be found in crystalline and amorphous semiconductors. The following kinds of defect can be found in conjugated polymers, and subsequently in small molecules:

– Point defects such as vacancies at lattice sites and interstitial atoms. Impurity atoms on interstitial and substitutional positions are also point defects.

– Extended defects such as dislocations, grain boundaries. In polymers, dislocations occur when the periodicity of the repeated units is broken in a given direction.

– Amorphous defects, which are disorders that change the lattice surrounding the defect site by forcing the neighboring atoms to move out of their normal position. These defects lead to random bonding between atoms with large deviations in the bond length and the bond angle, already seen in the description of the amorphous structure.

– Polymer defects, which are inherent to the structure of polymer materials. Strictly speaking, these defects can be included in the different previous defect kinds. However, they have specific characteristics and differ from the defect analog in inorganic solids. Among these, dangling chains and chain ends strongly influence the charge transport in a number of ways, for instance by acting as trap centers. Dangling chains can be compared to dangling bonds in amorphous semiconductors but as the polymer is mechanically flexible, the chains are not immobilized and can fit the structure of the material. On the other hand, chain ends whose density depends on the molecular weight of the polymer have been shown to have a significant influence on the structural ordering, charge transport and stability of materials (Martin et al. 1995; Choi et al. 2012). Chain ends are reactive sites and introduce defects into the polymers by disrupting the material order.

There are other possible types of defects in polymers such as fold surface and chain folds, which modify the structural order and introduce disorder to the polymer. Similarly, for both polymers and small molecules, the introduction of side groups may cause local distortions of the lattice by their steric hindrance.

2.4.4. *Defects in small-molecule crystals*

Small-molecule crystals are formed of molecules which are arranged in a herringbone packing with one or several molecules per unit cell. The structure of small-molecule films depends not only on the nature of the molecules but also on the nature of the substrate on which the film is deposited. Indeed, the interactions between the organic molecules and the atoms of the substrate surface strongly influence the organization and the structure of small-molecule thin films.

For a strong interaction, the contact between the two layers can be developed on a large area and the long axis of the molecules is parallel to the substrate surface. The lateral alignment of molecular layers on the substrate surface can take place, providing there is a local ordering under favorable growth conditions. However, because of the rigid bonds developed by the interactions, there is a large lattice mismatch resulting in a pronounced and local islanding of the grown organic films. For a weak interaction, the deposited molecules tend to form "standing-up" layers having a structure which is similar to that of the bulk. During their growth, local defects are created by the sliding of the molecules along their molecular axis (Kang et al. 2005). The orientation of the molecules on the substrate is generally random leading to a poor lateral alignment of molecular layers and consequently favors the formation of extended defects (grain boundaries) (Götzen et al. 2010).

Other point defects such as vacancies and impurities are also present in small molecules and are introduced to the materials during their synthesis, processing or by contact with the environmental media. Oxygen, water and hydrogen are frequently identified as contaminant impurities in small-molecule films. These defects create traps in the band gap and affect the transport of charge carriers by reducing the charge mobility. The high mobility observed in several small-molecule films corresponds to a low density of defects, which could be obtained by improving the quality of the contact structure between the organic layer and the substrate surface, which enables local epitaxial growth of the organic layer.

2.4.5. *Localized states*

As previously explained, conjugated polymers are considered as mixtures of crystalline and amorphous phases; the proportion of the different phases depends on the nature of the material, and also the synthesis technique and the deposition conditions employed. Consequently, the energy states in the amorphous regions are localized and those of the crystalline part are extended. In addition, as the binding energy between molecules in organics are of the van der Waals interaction, the width of the HOMO and LUMO bands is small. It can be considered that all the states in the band gap are localized (Horowitz 2015).

The origin of the localization of wave functions in conjugated polymers is usually explained by the *conjugation break model* (Mladenovic and Vukmirovic 2015). According to this model, the wave functions are delocalized in parts of the polymer where the coupling between monomers is strong. However, because of the weak bonding between chains, the monomers can rotate about their axes under mechanical or thermal stresses, varying the torsion angle between them. The conjugation length will be broken when the torsion angle is higher than a critical value and the wave functions become localized. It is shown that the delocalization length is large at the band edges of conjugated polymers and becomes reduced and localized when the energy levels move away from them. Further consideration of the electrostatic disorder on the wave functions shows that the structure of the polymer affects the formation of the localized states. The electrostatic disorder is introduced by the charge transfer in the monomer, which behaves as a dipole in the polymer chains. Because these chains have random orientations in the polymer, the potential created by the monomer dipoles varies randomly within the material, introducing the electrostatic disorder. From this point of view, any spatial modification of the polymer structure may affect the potential of monomer dipoles and therefore can change the electrostatic order leading to the formation of localized states. Consequently, adding side chains into the backbone of the polymer chains tends to reduce the disorder because these side chains act as spacers between chains and mitigate the influence of the monomer dipoles. On the other hand, disorder on the main chains strongly influences the localization and affects both intrachain and interchain electronic couplings. This picture allows for explanation of the temperature behavior of ordered conjugated polymers. Indeed, because of the weak van der Waals bonding between polymer chains, the mechanical properties of the polymer change when heated (chain displacement, distortion of chain segments), creating a *thermal disorder*, which increases the disorder of the material.

2.4.6. *Density of states*

According to the description of the organic SC structure, the density of states of organic materials can be depicted as follows. The main energy levels in organic semiconductors are the HOMO (highest occupied molecular orbital) and LUMO (lowest unoccupied molecular orbital) levels, which define the band gap of the material. States above the LUMO and those below the HOMO levels are extended. Due to the presence of amorphous regions, band tails exist around the HOMO and LUMO levels, inside which localized states are formed. On the other hand, mobility edges (electron and hole) define the energy level at which the transition between extended and localized states occurs, and a mobility gap can be determined from these levels as for amorphous semiconductors.

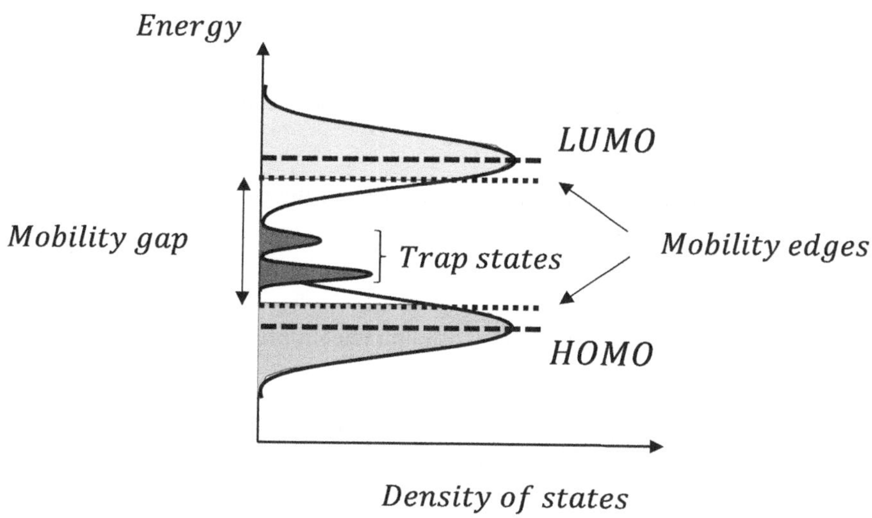

Figure 2.5. *Schematic representation of the DOS of organic semiconductors. For a color version of this figure, see www.iste.co.uk/nguyen/defects.zip*

The DOS of organic semiconductors is schematically represented in Figure 2.5 with the Gaussian form of the state distribution. The actual shape of the DOS is modified by the presence of defects, as discussed in section 2.4.3. These defects create localized levels inside the band gap of energy that can be either discrete or distributed. Depending on the nature of the defects, this level can be close to the HOMO or the LUMO or to the middle of the band gap. They are called *trap states*, which have a strong influence on the transport process in organic semiconductors, as will be discussed in the next chapter.

2.5. Distribution of the energetic states

The energetic states are distributed within the band gap of semiconductors and can be described by using two mathematical models: the Gaussian distribution and the exponential distribution.

The Gaussian distribution assumes a function of the form

$$g(E) = \frac{N_T}{\sqrt{2\pi\sigma^2}} \exp\left(-\frac{(E-E_T)^2}{2\sigma^2}\right) \quad [2.5]$$

where E is the energy of the state, E_T is the mean value of the distribution, N_T is the number of states per unit volume and σ is the width of the distribution.

The Gaussian distribution is experimentally observed in several physical characteristics of disordered materials, which describe the statistical variations of a mean value E_T by small amounts around it.

The exponential distribution assumes a function of the form

$$g(E) = \frac{N_T}{kT_0} \exp\left(\frac{E-E_T}{kT_0}\right) \quad [2.6]$$

where k is Boltzmann's constant, T_0 is a parameter representing the width of the exponential function.

The exponential distribution describes an exponential variation increase (or decrease) in the number of states towards (or from) a limit value E_T when the energy increases.

In Figure 2.6, the two types of distribution are represented with respect to the LUMO and HOMO levels of organic semiconductors. Although it is theoretically possible to distinguish between these distributions through the experimental determination of physical characteristics of organic devices, it is not easy to clearly identify the distribution type of the studied materials due to the large variation of experiment parameters (e.g. the temperature), which is not compatible with organic semiconductors.

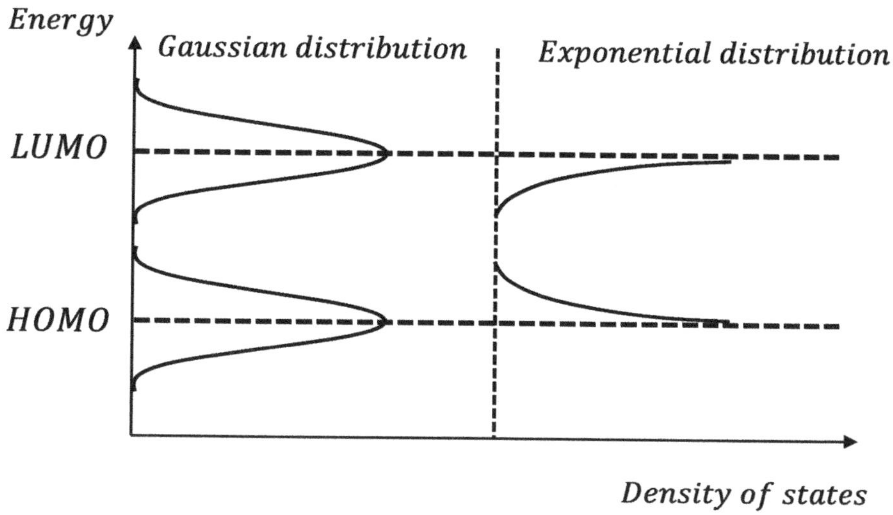

Figure 2.6. *Schematic representation of the distribution types of energetic states*

3

Defects and Physical Properties of Semiconductors

Optimal charge transport is achieved in crystalline semiconductors thanks to their highly ordered structure, in contrast with amorphous semiconductors, which are highly disordered. In intrinsic crystalline silicon (*Si*), the electron drift mobility is as high as 1,400 $cm^2V^{-1}s^{-1}$, while that of amorphous silicon is only 0.003 $cm^2V^{-1}s^{-1}$. The low carrier mobility, and hence the low electrical conductivity of the semiconductor, is mainly due to the disorder of the material, which is linked to the localized states previously described. Other physical properties of solids are affected by the disorder, and this will be the case for organic semiconductors, which are considered as paracrystalline materials.

This chapter will focus on the effect of disorder on the electrical and optical processes in organic semiconductors, since most of their applications such as OLEDs (Organic Light-Emitting Diodes), OPVs (Organic Photovoltaics) and OFETs (Organic Field-Effect Transistors) are built by exploiting the carrier electrical transport and the optical properties in the materials. The role played by other components such as metal electrodes or transport layers, which contribute to creating or increasing the disorder of the devices, will also be considered.

3.1. Carrier transport in organic semiconductors

Understanding the charge carrier transport in organic semiconductors is of primary importance in order to improve the carrier density in devices, and hence, their energetic performance characteristic.

In crystalline solids, the charge transport is characterized by the delocalized electrons whose displacement can be described by the wave functions according to Bloch's theory. Analysis of the carrier motion in a periodic potential leads to the concept of energy bands and *band transport*, which describes the transport process in the solids.

Accordingly, the electrical conductivity σ of a crystalline metal (containing only electrons) can be written as

$$\sigma = \frac{n\,e^2\,\tau}{m} \qquad [3.1]$$

where n is the electron density, e is the electronic charge, τ is the relaxation time and m is the electron effective mass.

By introducing the electron mean free path l and its velocity v, we can write the conductivity as:

$$\sigma = \frac{n\,e^2\,l}{m\,v} \qquad [3.2]$$

The conductivity can be expressed as a function of carrier *mobility* μ, which relates its velocity v to the electric applied field F by:

$$v = \mu\,F \qquad [3.3]$$

The carrier mobility expresses their ability to move in the material.

For crystalline semiconductors, which contain electrons and holes, the mobility of holes μ_p, and that of electrons μ_n, are generally different. For *Si* and at T = 300 K, $\mu_n \sim 1{,}400$ cm^{2}V^{-1}s^{-1} and $\mu_p \sim 500$ cm^{2}V^{-1}s^{-1}.

For some organic crystals (such as naphthalene or rubrene), which exhibit a temperature dependence like that observed in crystalline semiconductors, that is, an increase in the conductivity with increasing temperature, the charge transport is termed *band-like transport* (Sakanoue and Sirringhaus 2010). For the majority of organic semiconductors, because of the weak interactions between molecules and the disorder induced by structure, the energy states are almost localized. Therefore, the basic transport mechanism in these semiconductors is the *hopping process* between localized states.

It should be noted that the charge carriers in inorganic or organic semiconductors are electrons and holes. Under certain conditions, an electron in a localized state can significantly distort its surroundings, creating a lattice deformation around the charge. This process can occur in organic semiconductors because the presence of a charge carrier in an energetic state induces a deformation of the lattice surrounding the carrier due to the weak van der Waals bonding between molecules or polymer chains. The carrier charge (hole or electron) and the associated lattice deformation is named a *polaron*, which is a specific charge state (hole polaron or electron polaron) in organic semiconductors, and in particular, in conjugated polymers. A polaron is considered as a quasi-particle composed of its charge and the surrounding polarization cloud (Figure 3.1). The formation of polarons in disordered systems leads to a strong charge–phonon coupling and affects the transport process in these systems (Bässler et al. 1994). For the sake of clarity, in the following, no distinction will be made between the charge carriers (electrons or holes) and polarons. In other words, polaron relaxation to the equilibrium state after the charge displacement is ignored.

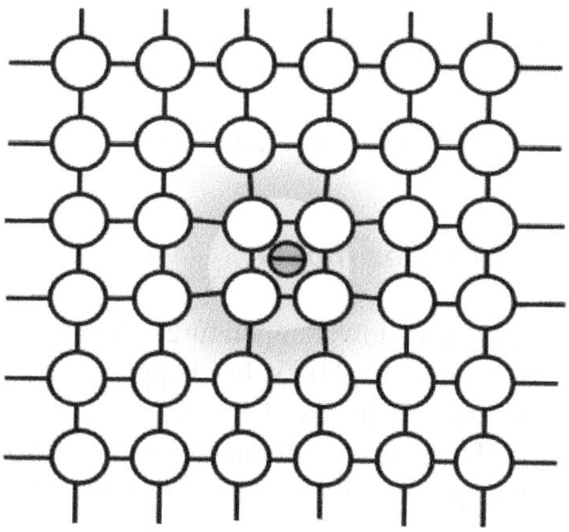

Figure 3.1. *Schematic representation of the formation of an electron polaron in the crystal lattice. For a color version of this figure, see www.iste.co.uk/nguyen/defects.zip*

3.1.1. *Hopping conduction*

By this process, an electron in an occupied state of energy below the Fermi level E_F moves to a nearby empty state of energy above E_F when it receives energy from a phonon. Here, the density of states is supposed to be uniform close to the Fermi level E_F. Although localized, an electron at energy E_i can tunnel to an adjacent state at energy E_j, since the energy difference is compensated for by the supplied energy, which enables the electron to overcome a potential barrier. This conduction process is called *phonon-assisted tunneling* and is described by Miller and Abrahams (1960) in doped inorganic semiconductors.

The transition rate for the hop from (occupied) site i to (empty) site j is given by:

$$\nu = \nu_0 \exp(-2\alpha R_{ij}) \exp\left(-\frac{E_j - E_i}{kT}\right), E_j - E_i > 0 \qquad [3.4a]$$

$$\nu = \nu_0 \exp(-2\alpha R_{ij}), E_j - E_i < 0 \qquad [3.4b]$$

where ν_0 is the attempt-to-jump frequency (~10^{12}–10^{13} Hz), α is the inverse localized length or *overlap factor*, and R_{ij} is the distance between sites i and j. The carrier hop depends on the distance separating two sites and the difference in energy between them. Hops from a high energy site to a low energy site depend only on the distance because the jumps are energetically favorable (non-activated process).

The electrical conductivity due to hopping is given by:

$$\sigma = \sigma_0 \exp\left(-\frac{\Delta E}{kT}\right) \sim \sigma_0 \exp\left(-\frac{1}{N(E_F)a^3kT}\right) \qquad [3.5]$$

where a is the distance between the nearest neighbors.

3.1.2. *Uniform density of states model*

According to Mott (1968), at low temperatures, the hopping process does not necessarily occur between nearest neighbors, but hopping probability between sites at bigger distances would be higher because the hopping carrier has a choice of many sites, and can find some with an energy similar to its own. This process needs only a weak energy supply and is expected to occur in solids with a high density of localized states. This hopping transport process is named *Variable Range Hopping* (VRH).

In Mott's description, the density of states is supposed to be uniformly distributed over an energy range and the hopping path of the electrons depends only on the temperature. Following Mott and David's analysis (1979), the electrical conductivity of a solid is given by the following expression (Mott's law):

$$\sigma(T) = \sigma_0 \left(\frac{1}{T}\right)^s exp\left\{-\left(\frac{T_0}{T}\right)^p\right\} \qquad [3.6]$$

Mott's characteristic temperature T_0 is given by:

$$T_0 \sim \frac{18}{kN(E_F)a^3(N)} \qquad [3.7]$$

where $N(E_F)$ is the density of states in the vicinity of the Fermi level and is supposed to be uniform.

The exponent s in the prefactor depends on the model and the semiconductor ($s = -\frac{1}{4}$ for Si and Ge). The exponent p in this expression is $p = \frac{1}{4}$, but will take the value of $p = \frac{1}{2}$ when there is a strong Coulombic interaction of electrons in the material (Efros and Shklovskii 1975).

There are some difficulties in exploiting Mott's law for organic semiconductors, mostly because the exponent p, which is obtained from the experimental data, cannot be determined very precisely, and therefore does not allow us to distinguish between the two values $p = \frac{1}{4}$ and $p = \frac{1}{2}$. There have been numerous experimental studies on common inorganic semiconductors, which supported either Mott's law with $p = \frac{1}{4}$ or Efros and Shklovskii's law with $p = \frac{1}{2}$. Furthermore, the hypothesis of a uniform density of states used for this analysis would not be appropriate for studying organic semiconductors, since these materials contain a high density of localized states, which are distributed in their band gap.

To consider the density of states in organic semiconductors and its influence on the conductivity of these materials, new investigations on the transport process have been proposed, which consider the hopping between localized states and at the same time, take the effects of both spatial and energy disorder into account.

3.1.3. *Non-uniform density of states models*

Because of their disordered structure, amorphous and organic semiconductors have non-uniform density of states, which are distributed in the band gap and can be described by either a Gaussian or an exponential function (equations [2.5] and [2.6]).

Inside disordered semiconductors, hopping of charge carriers between localized states is the main transport process. As already highlighted, in organic materials, energetic disorder is induced by the chemical structure of molecular groups contained in the volume of the sample (side groups or functional groups), and is therefore affected by the temperature. In contrast, spatial disorder is induced by the variations of the molecular orbital overlap, and is therefore temperature independent. Consequently, any transport parameter in organic semiconductors can be considered as a combination of two terms: one dependent and another independent of the temperature. This consideration allows the hopping process to be adapted to different configurations of the density of states to study the conductivity of the semiconductors.

3.1.3.1. *The Gaussian disorder model (GDM)*

In this model, developed first by Bässler (1993), the transport of injected carriers under an applied field F, is supposed to occur by hopping between localized sites with a rate given by the Miller–Abrahams model (equations [3.4a] and [3.4b]). The DOS is supposed to be distributed following a Gaussian function (equation [2.5]), in which the energetic disorder (or *diagonal disorder*) is expressed by the parameter $\hat{\sigma} = \sigma/kT$, where σ is the width of the distribution.

The hopping rate of a carrier between two sites i and j is expressed by:

$$\nu_{ij} = \nu_0 \exp(-2\alpha R_{ij}) \exp\left(-\frac{E_j - E_i - |e|FR_{ij}}{kT}\right),\ E_j - E_i - |e|FR_{ij} > 0 \quad [3.8a]$$

$$\nu = \nu_0 \exp(-2\alpha R_{ij}),\ E_j - E_i - |e|FR_{ij} < 0 \quad [3.8b]$$

The configurational disorder is induced by the variations of the distance between the lattice sites, which lead to a variation of the molecular overlap. The disorder parameter is called the *off-diagonal disorder* and is defined by: $\Sigma = 2\alpha R_{ij}$.

Using numerical Monte Carlo simulations for a cubic lattice of constant a, Bässler derived the conductivity of organic semiconductors under an electric field F and at temperature T by the following expressions:

$$\mu = \mu_0 \exp\left\{-\left(\frac{2}{3}\frac{\sigma}{kT}\right)^2\right\} \exp\left\{C\left[\left(\frac{\sigma}{kT}\right)^2 - \Sigma^2\right] F^{1/2}\right\},\ \Sigma > 1.5 \quad [3.9a]$$

$$\mu = \mu_0 \exp\left\{-\left(\frac{2}{3}\frac{\sigma}{kT}\right)^2\right\} \exp\left\{C\left[\left(\frac{\sigma}{kT}\right)^2 - 2.25\right] F^{1/2}\right\},\ \Sigma < 1.5 \quad [3.9b]$$

C is an empirical constant and is equal to $C = 2.9 \times 10^{-4} cm^{1/2} V^{1/2}$.

Introducing the parameter $T_0 = \frac{3\sigma}{2k}$, we can write the equations [3.9] as:

$$\mu = \mu_0 \exp\left\{-\left(\frac{T_0}{T}\right)^2\right\} \exp\left\{C\left[\left(\frac{3T_0}{2T}\right)^2 - \Sigma^2\right] F^{1/2}\right\},\ \Sigma \geq 1.5 \quad [3.10a]$$

$$\mu = \mu_0 \exp\left\{-\left(\frac{T_0}{T}\right)^2\right\} \exp\left\{C\left[\left(\frac{3T_0}{2T}\right)^2 - 2.25\right] F^{1/2}\right\},\ \Sigma < 1.5 \quad [3.10b]$$

The charge mobility is proportional to the electric field by $\mu \propto \exp(F^{1/2})$, which is known as a Poole–Frenkel behavior and is experimentally observed at high field in organic semiconductor devices. On the contrary, the carrier mobility depends on the temperature by the expression $\ln\mu \propto T^{-2}$. Nevertheless, the Arrhenius temperature dependence $\ln\mu \propto T^{-1}$ is also often observed and may be explained by the fact that the variations of the carrier mobility are small over the experimental temperature range, and both types of variations can be verified in the plot of the mobility function ln μ (T).

3.1.3.2. *Alternative Gaussian disorder models*

In the GDM model, the correlation between the site energies is not considered, that is, a carrier can jump to any empty site in its vicinity. This implies that the carrier density should be low in the semiconductor. Indeed, it is not and there are, in addition, a large number of occupied sites such as traps or impurities and consequently, the deviations from GDM are frequently observed.

An improved model with consideration of the influence of the charge density is proposed by Pasveer et al. (2005), who introduce the charge carrier density n as a parameter governing the transport. As a result, a fraction of the sites in the vicinity of the jumping carrier are occupied and not available for the jump. With this hypothesis, the *master equation* [3.11] is established and solved:

$$\frac{\delta}{\delta t}n_i(t) = \sum_{j\neq i} W_{ji}n_i(t)[1 - n_j(t)] + \sum_{j\neq i} W_{ij}n_j(t)[1 - n_i(t)] - \lambda_i n_i\,(t) \qquad [3.11]$$

According to the equation, the occupation probability $n_i(t)$ of an empty site i (first term) changes by the hop of a carrier from this site to another site j. Similarly, the occupation probability n_j of an empty site j (second term) changes by the hop of a carrier from this site to another site i. The third term expresses the decay of the charge density with recombination.

The mobility of the charge carriers, which depends on the energetic disorder, the temperature, the applied field and the charge concentration, is written as:

$$\mu(\hat{\sigma}, F, n, T) = \mu_0(\hat{\sigma}, T) \times \mu(n, \hat{\sigma}, T) \times f(F, \hat{\sigma}, T) \qquad [3.12]$$

with

$$\mu_0(\hat{\sigma}, T) = \mu_0 \exp(-C\hat{\sigma}^2) \qquad [3.13]$$

$$\mu(n, \hat{\sigma}, T) = \exp\left[\frac{1}{2}(\hat{\sigma}^2 - \hat{\sigma})\delta\right] \qquad [3.14]$$

$$f(F, \hat{\sigma}, T) = exp\left[0.44(\hat{\sigma}^{3/2} - 0.22) \times \left(\sqrt{1 + 0.8\left(\frac{F|e|a}{\sigma}\right)^2} - 1\right)\right] \qquad [3.15]$$

The parameter δ is given by:

$$\delta = \frac{2}{\hat{\sigma}^2}[\ln(\hat{\sigma}^2 - \hat{\sigma}) - \ln(ln4)] \qquad [3.16]$$

Good agreement with experiments performed in this study is obtained for the charge carrier density dependence of the mobility at various temperatures. In contrast, for the field dependence, large deviations are observed, showing the limits of the model.

3.1.3.3. *The transport energy level*

In the case of a Gaussian DOS with low carrier concentration n, most of the carriers are not localized near the Fermi level E_F at equilibrium, but around a level called *equilibration energy level* $E_\infty = -\frac{\sigma^2}{kT}$ (Baranovskii 2014). This level is usually represented by the average energy of the distribution. When the carrier concentration is small, the energy difference between the E_F and E_∞ is high ($E_F - E_\infty \gg kT$), so that the carriers tend to be localized near the equilibration level, rather than near the Fermi level. The charge transport occurs by activated hopping from the vicinity of the E_∞ level to a higher energy level called the *transport energy level* E_{Tr}.

As the Fermi level depends on the carrier concentration, when the carrier concentration increases, E_F increases and at a carrier concentration n_c, the two levels coincide, and the concentration can be determined by:

$$E_F(n_c) = E_\infty \quad [3.17]$$

When the carrier concentration is such that $E_F(n) > E_\infty$, the carrier hopping then occurs between the Fermi level and the transport energy level (Figure 3.2).

The transport energy plays the role of the effective energy to determine the activation energy of charge carriers. In organic and amorphous semiconductors, the carrier motion takes place in the upper tail region of the DOS, that is, the limit between the localized states and the extended states. Therefore, the transport levels correspond to the mobility edges in the DOS representation. From the states of energy lower than E_{Tr}, the carriers hop upwards with activation energy and from the states of energy higher than E_{Tr}, the hops are downwards in energy. If the electron transport level coincides with the electron mobility edge, then any electron occupying an energy state beyond E_{Tr} is mobile, and those with an energy below E_{Tr} are localized.

In simplified representations, the transport levels are, however, usually placed to coincide with the LUMO level for the transport of electrons and with the HOMO level for the transport of holes in organic semiconductors.

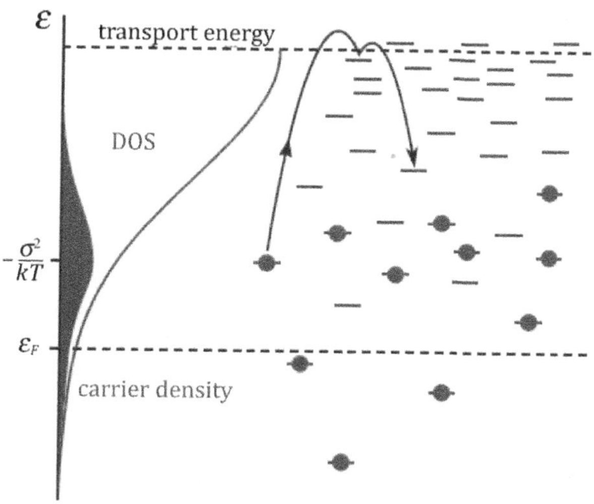

Figure 3.2. *Schematic equilibrium energy distribution of carriers in a Gaussian DOS (from Baranovskii (2014, p. 487)). For a color version of this figure, see www.iste.co.uk/nguyen/defects.zip*

3.1.3.4. *The exponential DOS model*

The exponential DOS is proposed by Vissenberg and Matters (1998) for studying the carrier mobility in OFETs by using the distribution function of equation [2.6]. The carrier transport occurs in the system by percolation through a network of sites, which are exponentially distributed. In systems with low charge densities and at low temperatures, the conductivity can be expressed by:

$$\sigma(n,T) = \sigma_0 \left\{ \frac{\pi\, n (T_0/T)^3}{(2\alpha)^3 B_C \Gamma(1 - T/T_0/) \Gamma(1 + T/T_0)} \right\}^{T_0/T} \qquad [3.18]$$

where $\Gamma(z) = \int_0^\infty \exp(-y) y^{z-1}\, dy$, and B_C is the critical number of connections of one site (B_C= 2.86).

The conductivity has an Arrhenius temperature dependence $\sigma \sim \exp\left(-\frac{E_A}{kT}\right)$, with an activation energy E_A depending on the temperature.

In systems with finite carrier concentration n, most of them occupy, in equilibrium conditions, the states around the Fermi level, and as the DOS decays exponentially from E_F, the density of carriers will decay in the same way (see Figure 3.2). Therefore, the transport will mainly depend on two energy levels – the transport energy and the Fermi energy, and occurs by activated hopping from the

Fermi level to the transport level, where states are available at high energies. It should be noted that, as the position of Fermi level E_F in the band gap depends on the temperature and the charge carrier concentration, the mobility in an exponential DOS model will be dependent on these variables, as previously mentioned.

3.2. Effects of defects on the carrier transport

The presence of defects in semiconductors affects the transport of charge carriers in different aspects. By breaking the lattice periodicity in crystals, defects change the potential distribution and create localized states in the band gap. Not only is the order of the crystal lattice affected, but the charge carrier conduction mechanisms are also modified when the defects density becomes strongly high, as in heavily doped semiconductors. Hopping of charge carriers is presumed to occur between localized states created by defects, and taking into account the structure of amorphous and organic semiconductors. The process applies to the materials used to study the charge motion. The disorder induced by defects changes the configuration of the DOS, which, in turn, influences the charge transport, as shown by the variations of the charge carrier mobility obtained by models or formalisms previously presented. By using the charge carrier mobility analysis, these models consider defects as a geometrical parameter, that is, their distribution and position (energetic or geometric) play a role in the motion of charge carriers, and consequently, affect their mobility.

There is another alternative to studying the effects of defects on the transport process in semiconductors. It consists of considering the electrical influence of defects on the carrier charges and the kinetics of these charges in interaction with defects. From this viewpoint, defects are electrically active and their interaction with mobile carriers will change the carrier density, which can be determined by measuring the current density. It is then possible to quantitatively estimate the density of defects in the semiconductor and devices, which allows for determination of the defect distribution and provides useful information on the DOS.

In the following, we give the description of electrically active defects or traps and their interaction with charge carriers created inside the semiconductors or injected from the electrodes under an applied electric field.

3.2.1. *Traps and recombination centers*

Traps and recombination centers can both be formed in the band gap of a semiconductor. The interaction processes with charge carriers are similar, but the effect results are different.

A *trap* is defined as a center that captures a free charge carrier, keeps it captured for a while, and then releases it when a sufficient amount of energy (thermal, electrical or optical) from an external source is provided to the carrier, allowing it to escape from the center. The process of carrier capture is called *trapping* and that of carrier release is called *detrapping*. Once released, the carrier returns to the nearest allowed band and moves freely here. A trap that captures electrons is called an *electron trap*. It is localized in the band gap and near the conduction band. A trap that captures holes is called a *hole trap*. It is localized in the band gap and near the valence band. The charge balance in the allowed bands is unchanged after the trapping and detrapping of charge carriers.

A *recombination center* is defined as a center that captures a free charge carrier, which will later recombine with a carrier of the opposite sign, resulting in a disappearance of both charges. The *recombination process* through recombination centers is similar to that occurring between the conduction and valence bands after the *generation* (or *excitation*) *process* by thermal excitation. In the latter, an electron in the conduction band recombines with a hole in the valence band, leading to the annihilation of both. The charge balance in the allowed bands is a loss of an electron and a hole after a recombination of charge carriers (Figure 3.3).

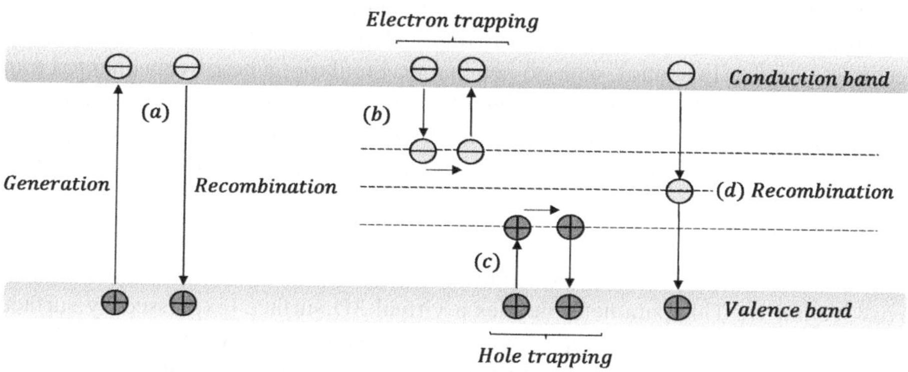

Figure 3.3. *Charge carrier transitions in semiconductors: (a) generation (excitation) and recombination by thermal excitation, (b) electron trapping by an electron trap, (c) hole trapping by a hole trap and (d) recombination by a recombination center. For a color version of this figure, see www.iste.co.uk/nguyen/defects.zip*

The distinction between trapping and recombination centers is, however, subtle because some centers can act as a trap or as a recombination center, depending on the conditions in which the semiconductor is working. It is generally accepted that a center, which is located near an allowed band edge, will be more likely to act as a trap than as a recombination center because its emission to the allowed band would require little energy to be operated. In contrast, a center lying near the middle of the band gap will probably act as a recombination center because the energetic conditions are favorable for the transitions between the center and the conduction and valence bands.

3.2.2. *Trapping mechanisms and trap parameters*

A charge carrier is captured by a trap center when it comes near the trap at a distance short enough that enables Coulombic attraction between the carrier and the charged trap. The Coulomb potential created by the trap center is:

$$V(r) = \frac{|e|^2}{4\pi\varepsilon_r\,\varepsilon_0 r} \qquad [3.19]$$

where ε_r and ε_0 are the relative and vacuum permittivity, respectively, and r is the distance between the charge carrier and the trap center.

Depending on the charge state of a trap, electrons or holes can be captured or not by electron or hole traps. There are three charge states of trap: positive, neutral and negative states. A positively charged trap state interacts with electrons by Coulombic attraction, and is therefore an electron trap. After the electrons capture, the trap becomes electrically neutral, with no net charge. Similarly, a negatively charged trap state is a hole trap and a neutrally charged trap state can be an electron or a hole trap. After the capture of a carrier, the neutral trap becomes charged by the captured carrier, that is, negatively with the capture of an electron and positively with a hole.

3.2.2.1. *Capture cross-section*

The probability for a trap to capture a charge carrier depends on its *capture cross-section* σ. This parameter defines a virtual 3D surface (expressed by surface unit cm^2) surrounding the trap center, inside which a moving carrier will be captured. For an electron trap, the number of electrons of density n and of velocity v that will be captured by the trap center during a time dt is comprised in a volume of space given by $\sigma_n vdt$. Therefore, the number of captured electrons is:

$$dn = nN_{tn}\sigma_n vdt \qquad [3.20]$$

The electron capture rate by electron traps is defined as the rate at which carriers are captured, and the variation of the electron density with time is expressed by:

$$c_n = -\frac{dn}{dt} = -C_n nN_{tn} \qquad [3.21]$$

where C_n is the *electron capture rate constant* or *electron capture coefficient*, and N_{tn} is the density of empty electron traps. From equations [3.20] and [3.21], we have:

$$C_n = v\sigma_n \qquad [3.22]$$

Taking the thermal velocity of the carriers for electrons, equation [3.22] becomes:

$$C_n = v_{th}\sigma_n = \left(\frac{3kT}{m_n^*}\right)^{1/2} \sigma_n \qquad [3.23]$$

where m_n^* is the effective mass of electrons.

The hole capture process is characterized by the *hole capture rate constant* or *hole capture coefficient* C_p given by:

$$C_p = v_{th}\sigma_p = \left(\frac{3kT}{m_p^*}\right)^{1/2} \sigma_p \qquad [3.24]$$

where m_p^* is the effective mass of holes.

The capture cross-section depends on the Coulombic interactions between the charge carrier and the trapping center, that is, the charge state of the traps and the distance between the carrier and the center. For electrons or holes, it can be determined by the physical characteristics of the materials.

Figure 3.4 shows the typical variation of the potential energy of trapping centers.

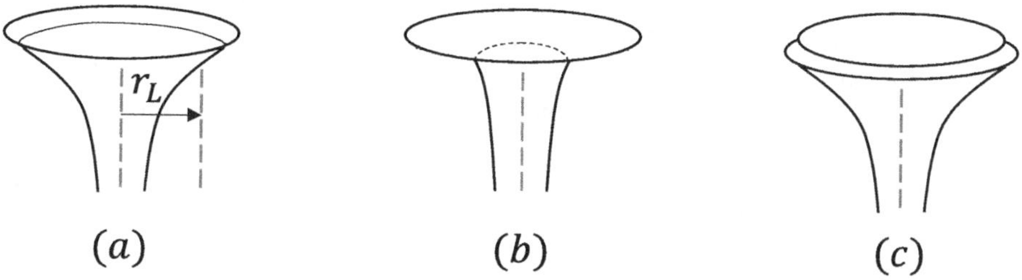

Figure 3.4. *Potential energy of: (a) an attractive trapping center, (b) a neutral trapping center and (c) a repulsive trapping center*

– An attractive center has a strong interaction with carriers and can catch a charge carrier in the vicinity of the center. To enable the carrier capture, the velocity

v_a of a carrier induced by the trap attraction at distance r should be higher than the diffusion velocity v_d of the same carrier.

$$v_a = \mu F = \mu \times \frac{|e|}{4\pi\varepsilon r} = \frac{|e|\tau}{m^*} \times \frac{|e|}{4\pi\varepsilon r} = \frac{|e|^2\tau}{4\pi\varepsilon m^* r} \quad [3.25]$$

$$v_d = \frac{2\lambda}{3r} v_{th} \quad [3.26]$$

where λ is the carrier mean free path, and τ is the carrier lifetime.

The critical distance r_L between the carrier and the trapping center will then satisfy the following condition:

$$r_L = \frac{3|e|\mu}{4\pi\varepsilon \times 2\lambda v_{th}} \quad [3.27]$$

The capture cross-section for an attractive trapping center can be expressed by

$$\sigma_{att} = \pi r_L^2 = \frac{9\pi|e|^4}{(4\pi\varepsilon)^2 \times 16(kT)^2} \quad [3.28]$$

For a semiconductor of relative permittivity of $\varepsilon_r = 3$, the calculated value of the capture cross-section at T = 300 K will be σ_{att} ~3.8 10^{-11} cm^2. Equation [3.28] is established on the hypothesis that collisions between the charge carriers are numerous within the capture cross-section of the trapping center and implicitly implies the condition for the mean free path of carriers $\lambda < r_L$. For many semiconductors, this condition is not fulfilled, and deviations from the calculated value of σ_{att} are experimentally observed.

– A neutral center has no charge and interacts with a carrier by an attractive force produced on the carrier, which results in its capture. A carrier of charge $|e|$ at a distance r from the center creates a dipole moment equal to $\alpha|e|/4\pi\varepsilon r^2$, and α is the polarizability. The attractive force on the carrier due to the dipole is then:

$$F_{att} = \frac{\alpha|e|}{4\pi\varepsilon r^2} \times \frac{|e|}{4\pi\varepsilon r^3} = \frac{\alpha|e|^2}{(4\pi\varepsilon)^2 r^5} \quad [3.29]$$

Here, the carrier capture depends on the polarizability of the material, and the capture cross-section is smaller than that of an attractive center.

– A center is repulsive for carriers of a given sign when it has captured a carrier of the same sign previously. For example, a center that has captured an electron becomes electron repulsive, that is, the probability for an electron approaching the center to be captured is low, since it has to overcome a potential barrier to reach the center. The capture cross-section of such centers is small.

It should be noted that the experimental values of capture cross-section determined by experiments can be high ($\sigma \sim 10^{-15}\, cm^2$ or larger) or low ($\sigma \sim 10^{-21}\, cm^2$ or smaller) (Mitonneau et al. 1979), depending on the nature of the centers and the materials. It should not be considered as a geometric interpretation because the right meaning of the capture cross-section is the ability of the center to interact with the moving carriers. A low value of σ simply indicates a low probability for the center to capture a carrier and should not be interpreted as a real surface of unrealistic value.

3.2.2.2. *Trap activation energy or trap depth*

The trap activation energy E_A is defined as the energy needed to release the captured carrier from the trap to the allowed energy band. For electrons and holes, the activation energy is given by:

$$E_{An} = E_C - E_{tn} \quad [3.30a]$$

$$E_{Ap} = E_{tp} - E_V \quad [3.30b]$$

where E_{tn} and E_{tp} are the energy levels of the electron and hole trapping centers in the band gap. Here, it is assumed that the traps are confined in a discrete energy level. It should be kept in mind that in reality and as previously explained, the spatial surroundings of a defect in a solid are inhomogeneous and may have different configurations. The use of a continuous distribution of traps of an exponential form or a Gaussian form is better to describe the trap states with a maximum trap density at the energy level E_t. For the sake of clarity and simplicity, in the following, we ignore the effect of distribution of traps and the trap activation energy will be defined by equations [3.30a] and [3.30b] for discrete or distributed traps.

Depending on the position of the trap level E_t with regard to the allowed band edges, traps can be classified into *shallow* or *deep traps*. A *shallow (deep) electron* trap is related to a trap level E_{tn} that is located above (below) the quasi-Fermi level E_{Fn} and *a shallow (deep) hole trap* is related to a trap level E_{tp} that is located below (above) the quasi-Fermi level E_{Fp}. The probability of capture of an electron (or a hole) is given by the Fermi–Dirac statistics:

$$f_n(E) = \frac{1}{1+\exp\left(\frac{E-E_{Fn}}{kT}\right)} \quad [3.31a]$$

$$f_p(E) = \frac{1}{1+\exp\left(\frac{E_{Fp}-E}{kT}\right)} \quad [3.31b]$$

For electrons, if $E_{tn} - E_{Fn} \gg kT$, then $f_n(E) \ll 1$, that is, most of the shallow traps are empty. If $E_{fn} - E_{tn} \gg kT$, then $f_n(E) \to 1$, that is, most of the deep traps are filled. The same results are obtained for shallow and deep hole traps. In other words, a shallow trap can be defined as a trap where a charge carrier is captured for a short time and is quickly released, leaving the trap unoccupied, while in a deep trap, the captured charge will stay a for a long time before its release. It should be noted that a released charge carrier can be trapped and released several times by the shallow traps, while the release times for deep traps can exceed the time scale of the experiment. Other definitions of shallow and deep traps have been proposed using the value of the trap activation energy as reference for comparison, although this value does not reflect any particular physical process and is usually chosen arbitrarily.

The release of a trapped carrier from the trapping center is assumed to be described by the Arrhenius relation, and the release probability per unit time is given by:

$$p = \nu \, exp\left(-\frac{E_A}{kT}\right) \qquad [3.32]$$

where ν is the *frequency factor* or *attempt-to-escape* of value in the order of the lattice vibration frequency (10^{12}-10^{14} Hz).

Let us consider the capture and release of electrons by and from a single electron trapping center. Let n be the concentration of free electrons in the conduction band, and N_t, n_t be the concentrations of the total traps and the trapped electrons (occupied traps).

According to Hall (1952) and Shockley and Read (1952), the electron capture rate (in $cm^{-3}s^{-1}$) by the traps is a function of the electron density in the conduction band, the capture cross-section and the density of empty traps. It is given by:

$$c_n = v_n \sigma_n n N_t [1 - f_t] = C_n n N_t [1 - f_t] \qquad [3.33]$$

where f_t is the probability that the trap level is occupied by an electron:

$$f_t = \frac{1}{1+\exp\left(\frac{E_t - E_F}{kT}\right)} \qquad [3.34]$$

$(1 - f_t)$ is then the probability that the trap state is empty, and thus capable of capturing an electron.

The electron emission rate (in s^{-1}) from the traps into the conduction band is a function of the density of filled traps and is given by:

$$e_n = E_n N_t f_t \qquad [3.35]$$

where E_n is the electron emission rate constant or the electron emission coefficient corresponding to C_n.

Similarly, the capture rate for holes from the valence band is:

$$c_p = v_p \sigma_p p N_t f_t = C_p p N_t f_t \qquad [3.36]$$

The hole emission rate from the traps into the valence band is given by:

$$e_p = E_p N_t [1 - f_t] \qquad [3.37]$$

In thermal equilibrium and according to the *principle of detail balancing*, the rate of capture and the rate of emission of electrons and holes must be equal. Therefore, the net rate of capture, that is, the number of captures minus the number of emissions should be zero. We can write:

$$c_n = e_n \text{ and } c_p = e_p$$

Using the expression of the electron (hole) density in the conduction (valence) band:

$$n = N_C \exp\left(-\frac{E_C - E_F}{kT}\right) = n_i exp\left(-\frac{E_F - E_i}{kT}\right) \qquad [3.38a]$$

$$p = N_v \exp\left(-\frac{E_F - E_V}{kT}\right) = n_i exp\left(-\frac{E_i - E_V}{kT}\right) \qquad [3.38b]$$

We have:

$$e_n = v_n \sigma_n N_C \exp\left(-\frac{E_C - E_t}{kT}\right) \qquad [3.39a]$$

$$e_p = v_p \sigma_p N_V \exp\left(-\frac{E_t - E_V}{kT}\right) \qquad [3.39b]$$

The electron (hole) emission into the conduction (valence) band increases exponentially when the trap level is close to the band edges.

The temperature dependence of the emission rate can be determined by replacing the known expression of the parameters. For electrons:

$$v_n = \left(\frac{3kT}{m_n^*}\right)^{1/2}$$

$$N_C = 2\left(\frac{2\pi m_n^* kT}{h^2}\right)^{3/2}$$

The temperature dependence of the emission rate is then expressed by:

$$e_n = \sigma_n \Gamma\, T^2 \exp\left(-\frac{E_{nA}}{kT}\right) \qquad [3.40]$$

A plot of $ln(e_n/T^2)$ versus $1/T$ will be a straight line whose slope will provide the activation of the level $E_{nA} = E_C - E_t$. The prefactor Γ can also be exploited to provide the capture cross-section of the electron traps.

In non-equilibrium conditions, the probability that the trap level is occupied by an electron or a hole given by the Fermi–Dirac function of equation [3.34] is applied by replacing the Fermi level by the quasi-Fermi level E_{Fn} or E_{Fp}.

In many experimental techniques, the variation of the carrier density, via the capture and the emission processes, as a function of time, is determined. For an electron trapping process, the capture and emission parameters c_n and e_n, the variation of the density of occupied trap states as a function of time can be described by the equation:

$$\frac{dn_t(t)}{dt} = n(t)c_n - n_t(t)e_n \qquad [3.41]$$

In equilibrium, there is no net transfer of electrons. Therefore, we have:

$$n_{to}e_n = n_0 c_n \text{ or } \frac{e_n}{c_n} = \frac{n_0}{n_{to}} \qquad [3.42]$$

where n_0 and n_{to} are the free electron density and the trapped electron density at steady-state equilibrium, respectively.

It can be seen from equation [3.42], for a semiconductor with few free electrons in the conduction band and many trapping centers ($n_{to} \gg n_0$), the capture kinetics will be faster than that of emission.

Equation [3.41] can be rewritten as:

$$\frac{dn_t(t)}{dt} = \frac{n_{to}}{N_{tn}-n_{to}} e_n - \frac{n_t(t)}{N_{tn}-n_{to}} e_n \qquad [3.43]$$

where N_{tn} is the total density of centers at the level E_t in the band gap, n_t is the trapped-electron density and $(N_{tn} - n_t)$ is the density of empty centers.

The emission rate is an exponential function of time, and the time constant is defined as the *relaxation time* $\tau_n = \frac{1}{e_n}$ of the process. Determination of the relaxation time can give access to the emission rate, which, in turn, provides information on the activation energy.

3.2.2.3. *Demarcation levels and recombination*

Recombination of charge carriers in the band gap of a semiconductor occurs through the capture of free carriers in an allowed band at recombination centers, followed by the recombination of the captured charge with free carriers of opposite sign. This indirect process is different from the direct band-to-band recombination, which occurs by recombination of free carriers in conduction and valence bands.

As in the case of the trapping centers, we define the recombination parameters to be used in describing the process.

In a band-to-band recombination, the recombination rate is defined by:

$$R = -\frac{dn}{dt} = -\frac{dp}{dt} = C_r np \quad [3.44]$$

$C_r = v\sigma_R$ is the band-to-band recombination rate constant. The velocity v in this expression is the average velocity of an electron and a hole.

Let us consider a center of acceptor-type, which can capture electrons from the conduction band. If this center is a recombination center, the captured electrons will recombine with holes in the valence band and the rate of electron capture should be equal to the rate of hole capture. Therefore, let N_{rn} be the density of total recombination centers and n_{rn} be the density of the captured electrons. The recombination rate related to this center can be expressed as:

$$R_n = v_n\sigma_n n(N_{rn} - n_{rn}) = v_p\sigma_p p n_{rn} \quad [3.45a]$$

Similarly, the recombination rate for a donor type, which can capture holes from the valence band to recombine with electrons in the conduction band, can be written as:

$$R_p = v_p\sigma_p p(N_{rp} - n_{rp}) = v_n\sigma_n n n_{rp} \quad [3.45b]$$

where n_{rp} is the density of the captured holes.

As previously stated, an electron trapping center may act as a recombination center, depending on its position in the band gap. The *demarcation level* E_{Dn} is defined as an energy level for which the probability of emission of an electron to the conduction band is equal to the probability of capture of a hole from the valence band by the same electron (Rose 1963). In other words, the rate of emission of captured electrons to the conduction band is equal to the rate of recombination of these electrons with free holes in the valence band. We can write:

$$v_n \sigma_n n(N_{tn} - n_t) = n_t \nu_n exp\left(-\frac{E_C - E_t}{kT}\right) \quad [3.46]$$

where N_{tn} is the total density of centers at the level E_t in the band gap, n_t is the trapped-electron density and $(N_{tn} - n_t)$ is the density of the empty centers.

Replacing the parameters by their expressions:

$$n = N_C \exp\left(-\frac{E_C - E_{Fn}}{kT}\right) \quad [3.47]$$

$$n_t = N_{tn} \frac{1}{1+\exp\left(\frac{E_t - E_{Fn}}{kT}\right)} \quad [3.48]$$

$$N_{tn} - n_t = N_{tn} \frac{1}{1+\exp\left(\frac{E_{Fn} - E_t}{kT}\right)} \quad [3.49]$$

The attempt-to-escape ν_n determined from equation [3.44] is then:

$$\nu_n = N_C v_n \sigma_n \quad [3.50]$$

The electron emission rate of captured electron is:

$$e_n = n_t N_C v_n \sigma_n exp\left(-\frac{E_C - E_t}{kT}\right) \quad [3.51]$$

The recombination rate of the captured electron with free holes from the valence band i:

$$R_n = v_p \sigma_p p n_t \quad [3.52]$$

The demarcation level $E_t = E_{Dn}$ is determined by the condition $e_n = R_n$. We can write:

$$n_t N_C v_n \sigma_n exp\left(-\frac{E_C - E_t}{kT}\right) = v_p \sigma_p p n_t$$

We obtain:

$$E_{Dn} = E_{Fn} + kTln\left(\frac{p\sigma_p}{n\sigma_n}\right) \quad [3.53]$$

Similarly for the localized states near the valence band, the rate of hole emission to the valence band is set equal to the recombination rate of the captured holes with free electrons from the conduction band at the demarcation level $E_t = E_{Dp}$. We obtain:

$$E_{Dp} = E_{Fp} + kTln\left(\frac{p\sigma_p}{n\sigma_n}\right) \quad [3.54]$$

Depending on the position of the demarcation levels, the localized states can act as trapping centers or recombination centers. This behavior can be examined by comparing the levels E_{Dn} and E_{Dp} with the quasi-Fermi levels E_{Fn} and E_{Fp}, which replace the Fermi level in non-equilibrium conditions of the semiconductor (under optical or electrical excitations, for instance). When $n\sigma_n = p\sigma_p$, the demarcation levels coincide with the quasi-Fermi levels and the localized states located between E_{Dn} and the conduction band will act as electron traps, those located between E_{Dp} and the valence band as hole traps and those between E_{Dn} and E_{Dp} as recombination centers (Figure 3.5).

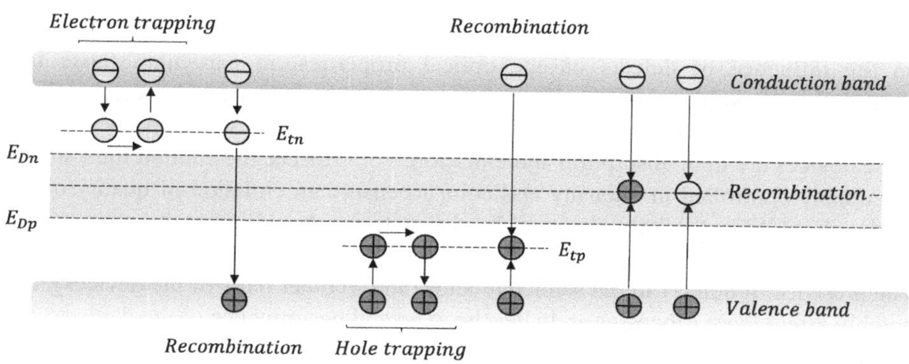

Figure 3.5. *Schematic representation of demarcation energy levels with* $n\sigma_n = p\sigma_p$. *For a color version of this figure, see www.iste.co.uk/nguyen/defects.zip*

In practice, the demarcation levels are not exploited for experiments, but allow for explaining the behavior of the carrier transitions between the localized states and the allowed energy bands.

3.2.2.4. *Organic semiconductors*

The mechanisms and parameters defined in the previous section apply to organic semiconductors, with a notable difference in the DOS structure of the materials, as already highlighted in Chapter 2. Indeed, the localized states in the band gap of organic semiconductors are not discrete, but distributed in energy in the band gap.

According to experimental data from the literature, Gaussian type trap distributions are likely adequate for the description of defects in organic semiconductors. Therefore, the trap parameters determined by analysis of a single-trap level model can be used by assuming that the single level coincides with the energy of the maximum of the distribution. Several experimental techniques used for trap measurements, such as the fractional thermally stimulated current, allow for the determination of the trap distribution profile with this assumption. It should be noted that the exponential distribution is also used for trap analysis, considering an exponential decay function starting from the conduction edge (E_C) or from the valence edge (E_V) into the band gap. Monte Carlo simulations of the charge transport in the presence of traps using both types of distribution give satisfactory results, and it seems that the type of trap distribution in the studied semiconductors would depend not only on their nature, but also on the story of the tested samples.

3.3. Optical properties of semiconductors and defects

There is a wide variety of optical processes in semiconductors, and we do not intend to examine all or many of them in this section. We will focus the discussion on the influence of defects on the optical properties, in particular, those that control the physical processes in organic semiconductor-based devices, such as light-emitting diodes or solar cells. Two optical key processes for the operation of these devices are light absorption and emission. For OLEDs, the emitted light should be suitably controlled in intensity and color to ensure an emission of quality of the diodes. For OSCs, the light absorbed by the material should cover the main part of the sunlight and should be high in intensity to ensure a good efficiency of the cells. The presence of defects in the semiconductor (and in other parts of the devices) may notably affect these processes and alter the expected performance of the devices.

3.3.1. *Defects and absorption*

3.3.1.1. *Absorption coefficient*

Light absorption by a semiconductor is expressed in terms of the absorption coefficient $\alpha(h\nu)$, which reflects the decrease in the light intensity during its propagation in the material.

The absorption coefficient depends on the difference in energy between the photon energy and the band gap of the semiconductor following the expression (Tauc et al. 1966):

$$\alpha(h\nu) = A^*(h\nu - E_G)^{1/n} \qquad [3.55]$$

In this expression, the prefactor A^* represents the slope of the plot of absorption coefficient versus the photon energy (or *Tauc plot*) in the linear region (see Figure 3.6). The exponent $1/n$ can take several values depending on the nature of the optical transition: allowed direct transition (1/2), allowed indirect transition (2), forbidden direct transition (3/2) and forbidden indirect transition (3). In most semiconductors, the transitions are direct and the plot of $\alpha(h\nu)$ presents a linear portion, whose slope is equal to 1/2 according to equation [3.55] (Figure 3.6a).

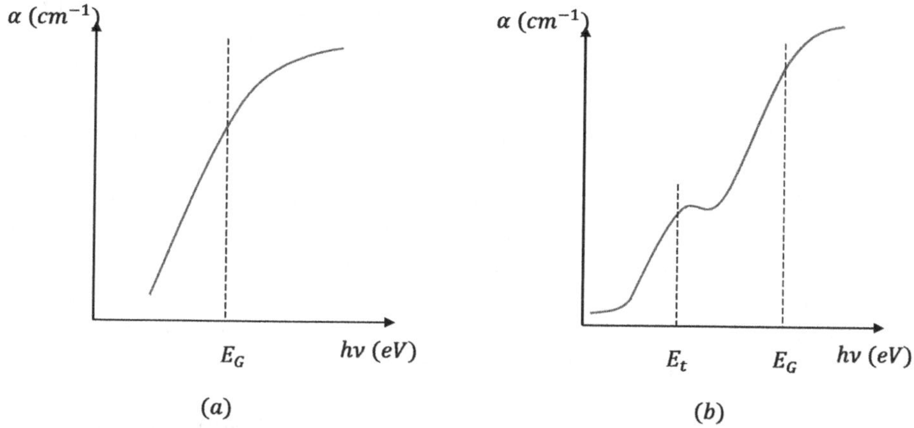

Figure 3.6. *Absorption edge of semiconductors with direct allowed transitions: (a) without defects and (b) with defects in the band gap. For a color version of this figure, see www.iste.co.uk/nguyen/defects.zip*

In an ideal semiconductor, absorption of a photon of energy $h\nu > E_G$ makes it possible to promote an electron from the valence band to the conduction band, leaving a hole in the former (*charge generation process*). Due to Coulomb attraction, the free carriers of opposite charges form an exciton that can move through the semiconductor. Such excitons are known as *free* or *Wannier excitons* because the electron and hole are separated by a distance higher than many lattice constants. The free electron–hole pair creates an attraction energy E_X called the *binding energy* and can be observed in the absorption spectra of semiconductors. According to equation [3.55], photons of energy $h\nu$ lower than E_G are not absorbed. However, because of the binding energy E_X, some discrete energies $h\nu < E_G$ can be detected in the absorption spectra, especially at low temperature conditions. The limit of the absorption line is then given by the expression:

$$h\nu = E_G - E_X \qquad [3.56]$$

When a free exciton comes to the vicinity of a lattice defect, especially impurity atoms, it can be trapped by the defects and loses its kinetic energy. Such an exciton is called *a bound exciton*. The loss of energy is E_{BX}, representing the binding energy of defect bound exciton.

3.3.1.2. *Absorption by trapping centers*

Consider a single electron trapping center close to the conduction band of a semiconductor. This center can capture an electron and becomes negatively charged. It can therefore attract a hole and form with it an exciton in a similar way as in the generation process. The absorption of such excitons can be seen in the spectrum by narrow bands or lines with variable intensity, which depends on the distance between the charge carriers. Indeed, the localized states are distributed in energy in the band gap of the semiconductor, and both electron and hole trapping centers are present. If the photon energy is higher than the *ionization energy* (or binding energy) of the trapping centers (electron or hole), absorption occurs and produces transitions from the centers to the conduction band, as schematically shown in Figure 3.7. The intensity of the absorption spectrum varies in the function of the photon energy and starts decreasing beyond the broad peak, due to the decrease in the probability of the transition from the trapping level to the conduction band.

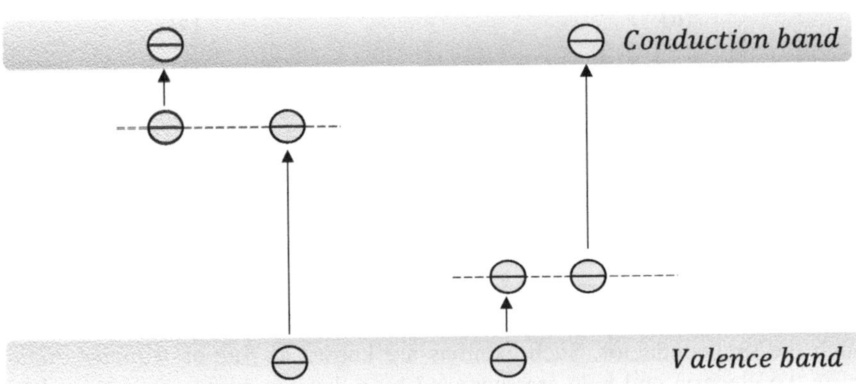

Figure 3.7. *Absorption transitions between trapping centers and the conduction band. For a color version of this figure, see www.iste.co.uk/nguyen/defects.zip*

Due to a high density of localized states on a trapping center level, the transitions between the trapping level and the conduction band constitute a broad absorption band in the spectrum. However, the intensity of the band is lower than that of the generation process because of the low density of localized states as compared with that of the valence band. Therefore, the absorption band due to the trapping centers appears as a shoulder in the absorption spectrum of the semiconductor (Figure 3.6b).

3.3.2. *Defects and luminescence*

In a perfect semiconductor, when the electrons are released from the conduction band to the valence band by the recombination process, the transition can be radiative or non-radiative. We have previously defined the emission and capture rates of electrons and holes in trapping process. For the luminescence of the semiconductors, we distinguish between two kinds of recombination rates depending on the nature of the transition.

A radiative transition occurs when the energy of the excited electrons transforms into photons after a certain time, called the *radiative recombination time* τ_r (or *radiative lifetime*). In contrast, in a non-radiative transition, the energy of the electrons is dissipated in the form of phonons (or heat) after a time τ_{nr}, called the *non-radiative recombination time* (or *non-radiative lifetime*). The probability of the transition per unit of time is equal to the inverse of the lifetime, and the total probability of transition can be written as:

$$\frac{1}{\tau} = \frac{1}{\tau_r} + \frac{1}{\tau_{nr}} \qquad [3.57]$$

A parameter, based on the carrier lifetimes and called the *quantum yield* or *quantum efficiency*, is defined to evaluate the quality of the luminescence process of the semiconductor. Its expression is:

$$\eta = \frac{1/\tau_r}{1/\tau_r + 1/\tau_{nr}} = \frac{1/\tau_r}{1/\tau} = \frac{\tau}{\tau_r} \qquad [3.58]$$

The quantum efficiency represents the ratio of the radiative recombination rate to the total recombination (radiative and non-radiative) rate and reflects the light emission ability of the semiconductor. A high value of η indicates that the (thermal, electrical or optical) energy received by the material is efficiently converted into photons and corresponds to a small loss of energy that is converted into phonons or heat.

In the generation–recombination process, if the transition is radiative, a photon of frequency ν is emitted, verifying the equation for free excitons:

$$h\nu = E_G - E_X \qquad [3.59]$$

The photon emission rate R_e depends on the density of carriers n in the conduction band, on the density of carriers p in the valence band and on the probability of radiative emission of the transition:

$$R_e = \beta np \qquad [3.60]$$

The parameter β is called the *bimolecular recombination coefficient.* Its value varies from $10^{-10}cm^3s^{-1}$ for direct band gap to $10^{-14}cm^3s^{-1}$ for indirect band gap semiconductors.

The recombination of bound excitons is characterized by emission lines at an energy lower than that of free exciton recombination, and the energy of the emitted photon is:

$$h\nu = E_G - E_X - E_{BX} \quad [3.61]$$

More generally, when a semiconductor that is an electron occupying a high energy level than that it would be under equilibrium conditions is released to an empty state of lower energy by a radiative transition, a photon of frequency ν_t is emitted, verifying the equation:

$$E_1 - E_2 = \Delta E = h\nu_t \quad [3.62]$$

In a semiconductor containing trapping centers (holes and electrons), the possible radiative transitions between carriers are indicated in Figure 3.8. They may involve one or two trapping levels and the carriers of the allowed bands.

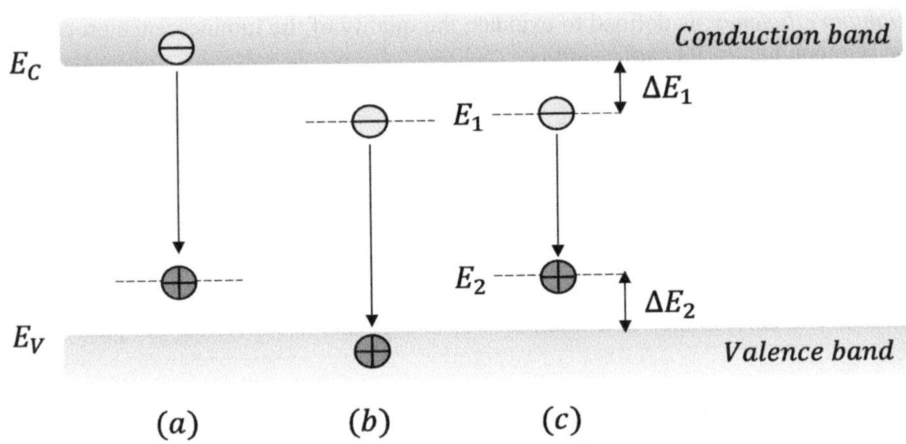

Figure 3.8. *Luminescence transitions between trapping centers and the allowed bands: (a) a hole trapping center and the conduction band, (b) an electron trapping center and the valence band and (c) a hole trapping center and an electron trapping center. For a color version of this figure, see www.iste.co.uk/nguyen/defects.zip*

3.3.2.1. *Free and bound excitons*

In a pure semiconductor, when the free excitons formed by electron–hole pairs recombine, the emitted spectrum is a narrow emission line of frequency verifying equation [3.59]. In the presence of defects, because of the localization of the bound excitons in the vicinity of these defects, there will be a loss of energy E_{BX} of the excitons due to the binding energy to the defect centers. Consequently, the photoluminescence spectrum of the semiconductor containing defects with radiative recombination will be red shifted, as compared with the spectrum of the pure materials. Furthermore, contrarily to free excitons, the spectrum shape of bound excitons is not sensitive to the temperature variation because of the lack of kinetic energy, that is, no broadening of the line shape with increasing temperature. It should be stressed here that the concentration of the bound excitons decreases with increasing temperature (*thermal quenching of luminescence)*, according to expression (Pankov 1971):

$$L(T) = \frac{L(0)}{1+C\exp\left(-\frac{E_i}{kT}\right)} \qquad [3.63]$$

where L(T) and L(0) are the emission intensity at the temperature T, T = 0 K, C is a temperature-independent constant describing the capture cross-section at the defect center and $E_i = E_X + E_b$ is the ionization energy of the bound exciton, with E_b being the additional energy binding the free exciton to the defect center.

A drop in luminescence intensity is observed when raising the temperature of the semiconductor, and the activation energy of the process can be determined using equation [3.63].

3.3.2.2. *Luminescence from trapping centers*

Carrier trapping is important in the luminescence of materials. The defects predominantly quench the luminescence of the semiconductor, but can also induce new channels of radiative combination.

In the presence of defects, the luminescence spectrum of a perfect semiconductor will show changes caused by the changes in the local polarization. In particular, the optical transitions between defect energy levels can induce changes in the emission frequency, according to equation [3.62].

In Figure 3.8, the energy of the emitted photon in the transition between an electron trapping and a hole trapping centers (process (c)) is:

$$E_1 - E_2 = E_G - (E_C - E_1) + (E_2 - E_V) = E_G - (\Delta E_1 + \Delta E_2) \qquad [3.64]$$

In this equation, $\Delta E_1 = E_C - E_1$ and $\Delta E_2 = E_C - E_2$ are the ionization energy of the electron and the hole trapping centers, respectively. In addition, the energy of the Coulomb attraction between the centers should be taken into account, and equation [3.64] becomes:

$$E_1 - E_2 = E_G - (\Delta E_1 + \Delta E_2) + \frac{(e)^2}{4\pi\varepsilon_0\varepsilon_r R^2} \quad [3.65]$$

where R is the distance between the centers.

As a result, the wavelengths of the luminescence emission are longer than those obtained from the generation–recombination process. Furthermore, as the difference in energy $E_1 - E_2$ depends on the distance R, which will have discrete values, the luminescence emission will have discrete lines or narrow bands, which correspond to the predominant Coulomb attraction (small values of R). As an example, the photoluminescence (PL) spectra of zinc oxide (ZnO) exhibit a large number of different bands in the visible region (Djurisic et al. 2010), which are assigned to the defect emissions. Indeed, the structure of the oxide is sensitive to the synthesis conditions, and the optical properties of the materials show large variations, depending on the fabrication technique. At room temperature, the PL emission lines can be found in a large range of frequencies from ~400 (blue) to ~750 nm (near-infrared), which are attributed to different types of defects that may be formed in the oxide: zinc interstitial, zinc vacancy, oxygen interstitial and oxygen vacancy.

It should be noted that recombination of carriers at trapping centers is not always radiative. In particular, surface defects or cluster defects usually form non-radiative centers and dissipate the excess energy by phonon emission. Recombination involving these defects does not produce light, but heat emission. The described luminescence process is also termed *recombination of donor–acceptor pairs*, as it results from the presence of both donors and acceptors in the semiconductor, which are ionized after the recombination.

Generally, the luminescence intensity emitted by a trapping center is related to the density of the corresponding traps.

3.3.2.3. *Kinetic of luminescence processes*

In the PL process, the luminescence intensity will decrease after removal of the optical excitation. The decay of the fluorescence with time depends on the nature of the involved carriers. There are two decay mechanisms: monomolecular and bimolecular recombination processes.

3.3.2.3.1. Monomolecular recombination

In this mechanism, only one molecule is involved, since the excited carrier will return to the ground state of the molecule after excitation. The recombination involves one free carrier at a time (e.g. an electron is captured by a center and then recombines with a hole). The decay rate is proportional to the density of the excited carriers. Let us consider excited electrons of density n, the decay can be described by the equation:

$$\frac{dn}{dt} = -\alpha n \qquad [3.66]$$

in which α is a constant.

The solution of equation [3.66] is:

$$n(t) = n(0)\, exp(-\alpha t) \qquad [3.67]$$

where $n(0)$ is the initial value of n.

The luminescence density can be written as:

$$i(t) = i(0)\, exp(-\alpha t) = \alpha n(t) \qquad [3.68]$$

3.3.2.3.2. Bimolecular recombination

In this mechanism, at least two free carriers are involved in the recombination (e.g. the generation–recombination process involves an electron of the conduction band and a hole of the valence band). The probability of recombination depends on the density of the carriers (e.g. electron density n) and the density p of available recombination centers. For the sake of simplicity, we suppose that $n = p$ (as in the case of the generation–recombination process). The decay can be described by equation:

$$\frac{dn}{dt} = -\beta np = -\beta n^2 \qquad [3.69]$$

in which β is the *bimolecular recombination coefficient* (from equation [3.60]).

The solution of equation [3.69] is:

$$n(t) = \frac{n(0)}{1+n(0)\beta t} \qquad [3.70]$$

The luminescence intensity can be written as:

$$i(t) = \beta n^2 = \frac{\beta n^2(0)}{[1+n(0)\beta t]^2} = \frac{i(0)}{[1+n(0)\beta t]^2} = \frac{i(0)}{\left[1+\sqrt{i(0)\beta t}\right]^2} \qquad [3.71]$$

The form of the luminescence decay function depends on the recombination mechanisms: for a monomolecular mechanism, $i(t)$ is an exponential function of time and for a bimolecular mechanism, it is a power–law function of time.

In both processes, as the initial luminescence intensity $i(0)$ depends linearly on the excitation intensity i_{ex} or the number of incident photons per surface unit and per unit time, the steady-state value of the luminescence depends linearly on the excitation intensity.

The luminescence response of a semiconductor excited by an excitation pulse of width t_C is schematically represented in Figure 3.9. It is supposed that t_C is long enough to allow the luminescence intensity to reach its steady-state value.

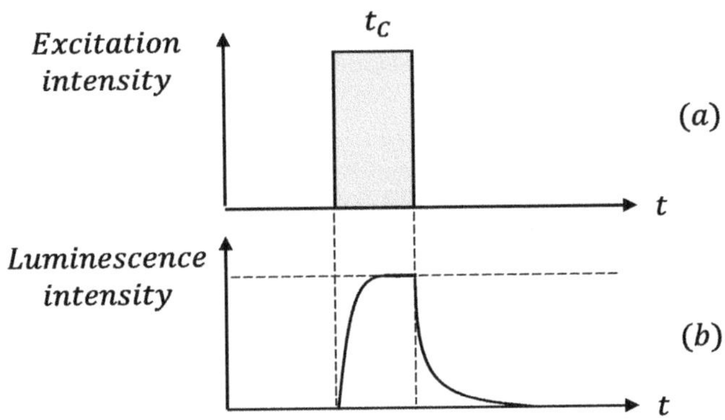

Figure 3.9. *Time dependence of the luminescence intensity under an applied light pulse: (a) light pulse of width t_C and (b) luminescence intensity growth and decay*

4

Techniques for Studying Defects in Semiconductors

Defects in semiconductors play a primary role in the physical processes which occur in the materials. Their effects on the properties of the semiconductors can be either beneficial (for instance, doping by impurities) or unfavorable (for instance, charge transport due to disorder). In conventional semiconductors, intensive investigations of defects have been carried out on a large panel of materials using techniques which have become increasingly sophisticated and developed thanks to the acquired knowledge and expertise on the defects. The techniques used to study defects cover a large physical field, exploiting the electrical, optical, magnetic and thermal properties of semiconductors. Most of these techniques have been successfully adapted to the study of organic materials, and despite the difference in some physical properties and processes of the semiconductors, the results obtained from defect determination in organic materials and devices have been coherent and reliable.

In this chapter, we will select and present the measurement techniques that are frequently used for studying defects in organic semiconductors. Some of these techniques are readily available in most laboratories; others need specific equipment and technical training but are still accessible. Using these techniques, the study of defects in organic semiconductors and devices has been investigated, and the results will be presented and commented on in the next chapter.

4.1. Electron spin resonance (ESR)

The ESR (or EPR – Electron Paramagnetic Resonance) technique has been widely used to study the defects on the atomic scale in silicon (Stesmans 1989) and other conventional semiconductors (Sudandana 1998). Recently, the technique has been applied to organic materials, and although it is still not widespread, several interesting results have been obtained, demonstrating how it is an efficient tool for studying organic semiconductors.

4.1.1. *Basic concepts of ESR*

ESR is used to study materials with unpaired electrons by applying an external magnetic field H_0 to the sample, which tends to align the magnetic moment vector μ of electrons with the direction of the field. Under the applied field, an electron precesses around the field direction at an angular frequency ω_0:

$$\omega_0 = \gamma \times H_0 \quad [4.1]$$

where γ is the magnetogyric ratio. For a free electron, $\gamma = 17.6 \times 10^6 sec^{-1} Gauss^{-1}$.

The energy of a magnetic moment μ placed in the magnetic field H_0 is:

$$E = \mu \times H_0 \cos\theta \quad [4.2]$$

where θ is the angle between the magnetic field and the direction of the dipole axis. For a free electron, there are two allowed values for the energy, which correspond to two spin quantum numbers $m_s = \pm\frac{1}{2}$. Therefore, the energy of the electron will be split into two levels separated by an energy of $2\mu_B H_0$. For a free electron in solids, because of additional interactions, the energy is expressed in function of the applied field by:

$$E = g\mu_B H_0 \quad [4.3]$$

where μ_B is the Bohr magneton, g is a factor (g or *Landé factor*) which is determined by measuring the field and the frequency at which resonance occurs. For free electrons, $g_e = 2.0023$. If the measured value of g deviates from g_e, the electron is supposed to gain or lose angular momentum through *spin–orbit coupling*, and the nature of the atomic orbital containing the unpaired electron can be investigated by the change in the g factor.

To detect the energy splitting of electrons under the applied magnetic field, an electromagnetic radiation or microwave beam of frequency ν is used to excite the transition by resonance. The frequency of the beam is such that:

$$h\nu = g\mu_B H_0 \quad [4.4]$$

In practice, the frequency of the microwave beam is kept constant, and the field is slowly swept through the resonance condition. The reflected power is measured providing an *absorption spectrum* or *ESR line* with finite width. The *first derivative* of the absorption spectrum is usually recorded in experiments and used for presentation of the ESR spectrum (Figure 4.1).

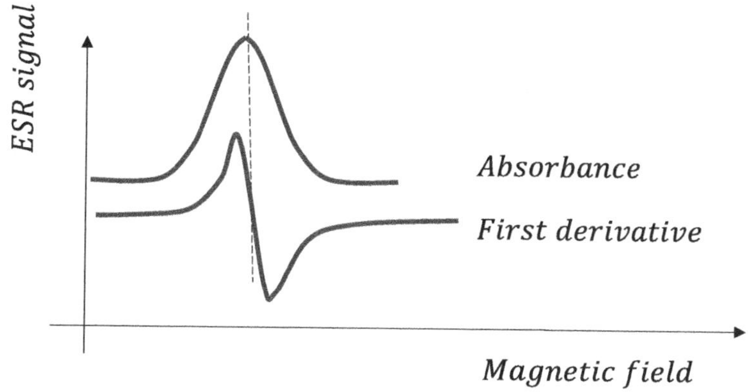

Figure 4.1. *ESR absorption spectrum and its first derivative*

4.1.2. *Interpretation of ESR line*

As the probability of occupation of the energy states is given by the Boltzmann distribution, the power absorbed by electrons is proportional to the total number of unpaired electrons, on the one hand, and is dependent of the temperature, on the other hand. Furthermore, as the energetic transitions occurs between electrons in lower energy level to higher energy level, when the temperature decreases, the power absorption increases providing better ESR lines. Integration of the spectrum over the magnetic field interval provides the total number of unpaired electrons.

The line shape of the ESR spectrum depends on the relaxation mechanism by which the electrons in the high energy state lose their energy to their neighboring environment. An electron relaxation can occur either by interaction with the lattice (*spin–lattice relaxation*) or by interaction with other centers such as defects (*spin–spin relaxation*). Both types of relaxation are characterized by a specific relaxation time, defined as the time for the system to lose $1/|e|$ of its excess energy. The relaxation time T_1 of the spin–lattice process depends on the materials. For instance, free radicals $\left(S=\frac{1}{2}\right)$ have a long relaxation time and a narrow ESR line, while transition metal ions have a short relaxation time and a wide ESR spectrum. When the spin couples with nearby nuclear spins (*hyperfine coupling*), additional energy states are introduced giving rise to additional lines in the ESR spectrum.

Under the applied field H, the nucleus can take either the configuration with the component of the nuclear spin quantum number $M_I = +1/2$ or the configuration with $M_I = -1/2$. The two applied fields at which resonance occurs are:

$$H = \frac{h\nu}{g\mu_B} = H_0 \pm H_e = H_0 \pm aM_I \quad [4.5]$$

where a is the hyperfine splitting constant.

Equation [4.4] can be written as:

$$h\nu = g\mu_B H_0 \pm \frac{1}{2} ha \quad [4.6]$$

Consequently, the ESR is split into two-line hyperfine structures with two transitions at H_1 and H_2.

The spacing between the lines reflects the degree of interaction between the unpaired electrons and the nuclei. The shape of the line can be usually described by a Gaussian function. The relaxation time T_2 of the spin–spin process is generally shorter than that of the spin–lattice process and the line shape is rather described by a Lorentzian function.

4.1.3. *Electron nuclear double resonance (ENDOR)*

ENDOR is an improved technique of the ESR for a two-spin system involving one electron $\left(S = \frac{1}{2}\right)$ and one proton $\left(I = \frac{1}{2}\right)$ interacting with a magnetic field. In other words, the technique allows for ESR and NMR (nuclear magnetic resonance) to be simultaneously realized for the system if it is possible to saturate partially the ESR signal.

Firstly, the ESR experiment is performed with the application of a constant magnetic field to the system with increasing microwave power until reaching a partial saturation of the transition, which increases the difference in occupation of nuclear spin levels. A radio frequency field is then applied to the system until reaching a frequency ν_1, for which the ESR signal starts to desaturate. The ENDOR transition will be observed when the mutual spins flips occur, that is, when $\Delta(M_s + m_I) = 0$. In this case, there are two transitions occurring at nuclear frequencies (*ENDOR frequencies*) ν_1 and ν_2 corresponding to the two orientations $m_s = \pm\frac{1}{2}$ of the electron spin. The NMR selection rule holds $(\Delta M_s = 0, \Delta m_I = \pm 1)$ and is opposite to the ESR selection rule $(\Delta M_s = \pm 1, \Delta m_I = 0)$. The ENDOR frequencies ν_1 and ν_2 verify the following relations:

$$h\nu_1 = -\frac{1}{2}ha + g_N\mu_N H_0$$

$$h\nu_2 = +\frac{1}{2}ha + g_N\mu_N H_0 \qquad [4.7]$$

where g_N and μ_N are the nuclear g-factor and the nuclear Bohr magneton respectively.

The analysis of the ENDOR spectra provides, on the one hand, the parameter a by measuring the difference of the ENDOR frequencies ν_1 and ν_2 and, on the other hand, the g_N factor by the determining mean frequency $(\nu_1 + \nu_2)/2$. These parameters enable us to identify the interacting nucleus.

4.1.4. *Investigation of defects using the ESR technique*

Analysis of the ESR spectrum recorded in a semiconductor can allow the determination of the basic parameters (g-factor, line width), which can be used to identify the paramagnetic states that give rise to the observed signal. To collect supplementary information of the defect states in the studied material, other defect measurement techniques are usually carried out under similar conditions to support the ESR result interpretation. Furthermore, by varying the experimental conditions used in ESR measurements such as doping and excitation, it is possible to establish the correlation between the resonance and the experimental parameters, and to interpret and explain the origin of the defects.

For the illustration of the measurements of defects by the ESR technique, results obtained in CdSe quantum dots (QDs) (Almeida et al. 2016) are presented here. QDs are often used as active material with or without conjugated polymers in organic devices such as OLEDs or OPVs. Figure 4.2 shows the spectra of QDs in the dark and under white light illumination.

The g-factor determined from the spectra is $g = 2.0045$, and is assigned to positively charged Se vacancies (V_{Se}^{+}), which are found to be localized in the bulk of the nanoparticles. From the quantification of the intensity of the spectrum, a defect density of $\sim 10^{19} cm^{-3}$ is determined. It is increased two-fold upon photoexcitation and suggests that another charge state exists, which cannot be detected when performing the ESR experiment in the dark. The density of these charge states is comparable to that of the Se vacancies. Further ESR investigations are carried out on silver-doped QDs, which are expected to be of N-type. By analyzing the intensity of the ESR signal obtained in samples with various Ag concentrations, we can conclude that the visible charge states (Se vacancies) act as electron traps whereas the invisible charge states act as hole traps.

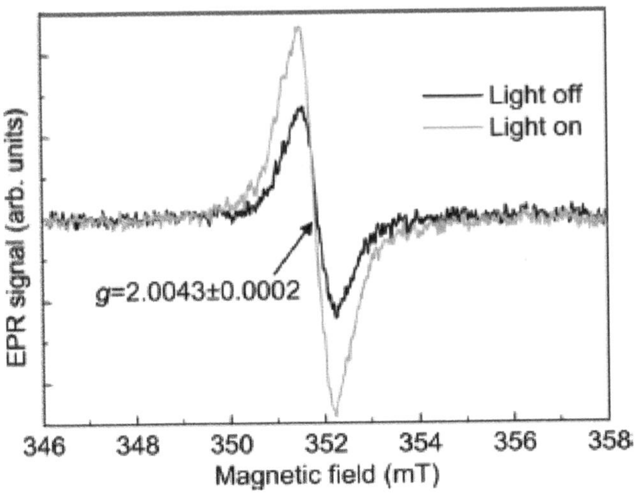

Figure 4.2. *Comparison between EPR spectra of undoped CdSe QDs measured at room temperature, in the dark (black line) and under white light illumination (yellow line) (from Almeida et al. 2016, p. 13763). For a color version of this figure, see www.iste.co.uk/nguyen/defects.zip*

In the organic electronic field, investigations of defects by the ESR technique have been performed on semiconductors and devices for studying defects created by doping and degradation process.

In the doping process of polymers, the increase of the conductivity is found to be linked to the enhancement of the spin density, which can often be related to the density of charges, and the charge carrier mobility in poly(vinylene disulfide) doped with protonic acid by ESR and SQUID (Superconducting Quantum Interference Device) measurements (Nguyen et al. 1994). Analysis of the results indicates that the increase in the conductivity of the doped sample is principally due to the improvement of the charge mobility because the magnetic susceptibility of the doped sample is independent of the temperature. The enhancement of ESR intensity is not directly responsible for the improvement of the conductivity but contributes to this increase by the charge transfer with spins between polymer chains (interchain transport).

Analysis using EPR technique of P3HT films doped with a strong Lewis acid BCF showed that the ESR hole densities correspond to the mobile hole densities determined from the electrical measurements with a *doping efficiency* (defined as the ratio of induced mobile charges to the amount of dopants) of ~18%. Furthermore, a linear relationship between the doping concentration and the spin density measured from the ESR spectra indicating that the dopants create mobile holes in the polymer, but the majority of P3HT cations are bound to the BCF anions and show no ESR activity (Pingel et al. 2016).

ESR measurements have been also performed to study degradation of organic semiconductors and devices. ESR spectra recorded in tris-8-(hydroxyquinoline) aluminum (Alq3) showed a g value of ~2.008 at room temperature. Its intensity increased linearly with the exposure time to humid air, but there is no shift in the ESR peak. The variation of the ESR signal was correlated with the decrease in the PL intensity with time and suggests the formation of free radicals in the degradation of the organic material (Roy et al. 2004).

In small-molecule CBP-based OLEDs, substantial losses of light emission over time were observed and were attributed to the formation of chemical products resulted from a chemical process of operational degradation (Kondakov et al. 2007). The degradation products were identified as BCP which forms radical species as detected by ESR measurements. The variation concentration of radicals as a function of irradiation time is, however, nonlinear, which suggests a possible accumulation of the free radicals during the operation of the device. The radicals act as deep traps (or non-radiative recombination centers) and also as luminescent-quenchers, and lead to the loss of light emission of devices.

The defects induced by aging and thermal annealing of regio-regular rrP3DDT:PCBM blend thin films were studied using a *light-induced electron spin resonance* (LESR) technique (Bonoldi et al. 2014). This technique allows for studying unpaired spins which are formed by the light excitation by comparing the ESR spectra recorded in the dark and after illumination. After switching off the excitation light, the decay of the ESR signal is monitored for obtaining the LESR recombination kinetics, which provide information of the trap characteristics of the materials. The ESR spectrum of the blend film in dark conditions shows the contribution of the polymer with a g value of ~2.002 and that of the fullerene with a g value of ~1.999. After aging for 21 days in static vacuum, the polymer signal P^+ is unchanged while the fullerene signal $PCBM^-$ decreases in intensity. After thermal annealing at 150°C for 20 min, the intensity of both signals decreases significantly. The results are explained by the contamination of the blend by residual oxygen during the aging period and the reduction of the deep trap density during the thermal treatment. The LESR intensity decreases after aging and annealing suggesting a loss of charge generation efficiency and an increase in phase separation. From the LESR recombination kinetics, the DOS calculation provides the distribution of the P^+ and the $PCBM^-$ trap states with peaks at 31, 75 and 36 meV (fresh, aged and annealed polymer) and 59, 101 and 49 meV (fresh, aged and annealed PCBM).

The ESR technique has been used for studying the charge formation and trapping in pentacene-based solar cells of structure $ITO/(PEDOT\colon PSS)/pentacene/C_{60}/BCP/Al$ (Marumoto et al. 2012). Here, PEDOT:PSS is poly(3,4-ethylenedioxythiophene):polystyrene sulfonate and BCP is bathocuproine. Without the PEDOT:PSS layer, the ESR signal is negligible and its intensity strongly increases, with a g value of ~2.002 when the PEDOT:PSS is incorporated. This observation suggests that spin species whose number is 4.0×10^{12} are formed at the PEDOT:PSS/pentacene interface. Under light irradiation of the device, the solar performance is degraded and the ESR intensity gradually decreases with the exposure time and its line width increases. The changes of the ESR characteristics indicate a chemical degradation of the organic material and possibly a reaction of radicals with residual oxygen at the PEDOT:PSS/pentacene interface may also occur.

4.2. Optical techniques

Among numerous optical techniques used to study semiconductors, the luminescence technique provides useful information on the main point defects such as the nature, the concentrations and the energy levels.

4.2.1. *Fluorescence spectroscopy (FL)*

There are two types of measurement techniques: *steady-state* (SSPL) and *time-resolved photoluminescence* (TRPL). The steady-state measurements are performed with a constant illumination and the steady-state is established almost immediately, which enables the spectra to be recorded. The time-resolved measurements are performed by applying a pulse of light to the sample and by recording the intensity fluorescence decay as a function of time. With a single decay time or lifetime τ, the fluorescence intensity is given by:

$$I(t) = I_0 \exp(-t/\tau) \qquad [4.8]$$

where I_0 is the intensity at $t = 0$.

In the presence of two independent decay mechanisms, equation [4.8] becomes:

$$I(t) = I_1 \exp(-t/\tau_1) + I_2 \exp(-t/\tau_2) \qquad [4.9]$$

where I_1, and I_2, and τ_1, and τ_2 are the amplitudes and the lifetimes of the double-exponential decay.

More advanced technique such as *fluorescence lifetime imaging microscopy* (FLIM) can provide information on physical phenomena at the nanoscale, that is, emission occurring inside the material not only in steady state but also with a time-resolved possibility.

Investigations of defects by the fluorescence spectroscopy consists of measuring the FL spectra using steady-state and time-resolved techniques (SSPL and TRPL). Quantitative estimation of the defect concentration can be obtained by comparison of the intensity ratio between the defect emission and the near-band edge emission spectra. The nature of the defects can also be determined by analyzing the FL intensity as a function of temperature (Reshchikov and Morkoç 2005). It should be noted that the quantitative determination of defects by the technique is based on the hypothesis of the simple emission process from defects. In reality, the evaluation of the defect concentration may be complicated by the experimental conditions, and the obtained result can be considered as an estimation.

In organic semiconductors, the FL and absorption techniques are applied for studying defects in Alq3 small molecules (Aziz and Narasimhan 2002). It is demonstrated that the Alq3 absorption occurs between 3.18 and 2.5 eV. Below 2.5 eV, surface states of density of $\sim 2 \times 10^{15} cm^3$ are responsible for the excitons trapping process.

The study of the FL has been investigated in conjugated polymers with varying backbone conformation and at different temperatures (Guha et al. 2003). The PL spectra of the polymers show all three main vibronic peaks labeled as the 0-0, 0-1 and 0-2. The 0-0 transition has the highest energy, taking place between the 0th level in the excited state and the 0th level in the ground state. The 0-n transition takes place between the 0th level in the excited state and the nth level in the ground state and involves the creation of a phonon. With increasing temperatures, the PL spectra of conjugated polymers such as polyfluorene (PF) or methyl-substituted ladder-type poly para-phenylene (MeLPPP) are blue-shifted, explained by the localization of the excitons due to an increase of the disorder processes in the polymer. In PF, a broad peak centered at 2.3 eV is observed at a high temperature, which corresponds to the emission of the keto defects sites.

Defects in P3HT are investigated by FL spectroscopy using both steady-state and time-resolved PL measurements. The polymer used in blend with PCBM fullerene is promising for solar cell applications. In P3HT films, in addition to the disorder introduced by torsional defects and chemical defects, two kinds of packing (structural) defects exist: J-aggregation and H-aggregation, which result from the intrachain and interchain interactions respectively (Spano and Silva 2014). The disorder of π-stacking impacts on the carrier transport, which is governed by the hopping mechanism. The shape of the PL spectra reflects the nature of the aggregate disorder with the intensity ratio I^{0-0}/I^{0-1}, where I^{0-0} is the intensity of the vibronic transition 0-0 and I^{0-1} is the intensity of the sideband transition (0-1 or higher). With an increasing temperature or increasing disorder, the I^{0-0}/I^{0-1} ratio decreases in J-aggregates and increases in H-aggregates because the 0-0 transition intensity depends on the structural molecular orientations.

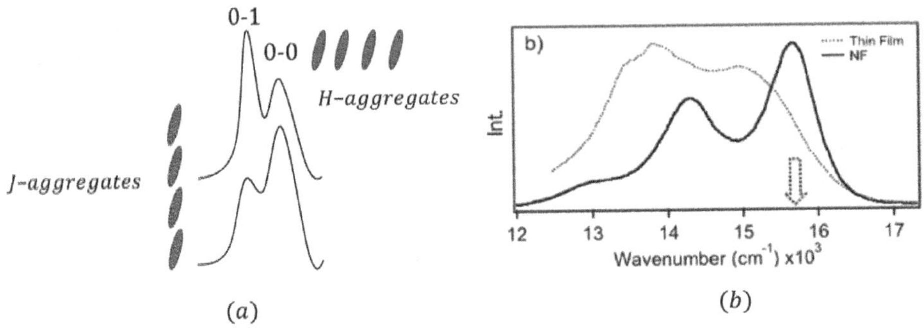

Figure 4.3. *Steady-state PL spectra in P3HT: (a) Schematic representation of 0-0 and 0-1 peaks with H-aggregates and J-aggregates, (b) emission spectra of P3HT thin films (blue dotted trace) and dilute NF (nanofibers) suspensions (black solid trace) (from Niles et al. 2012, p. 259). For a color version of this figure, see www.iste.co.uk/nguyen/defects.zip*

Nanofibers of P3HT diluted in toluene exhibit single-chain J-aggregate character (the I^{0-0}/I^{0-1} ratio is higher than unity) while P3HT thin films shows in contrast interchain coupling with H-aggregate character (the I^{0-0}/I^{0-1} ratio is less than unity) (Niles et al. 2012). Time-resolved photoluminescence results indicate two decay regimes (Labastide et al. 2012). In the short time regime ($< 5ns$), the decay follows an exponential law with biexciton and single radiation decay processes. In the long-time regime ($> 5ns$), the decay follows a power law, which is supposed to result from deep traps in the polymer.

In hybrid halide perovskites, the FL technique is used for characterizing the defects in several works. The carrier recombination dynamics in $CH_3NH_3PbI_3$ ($MAPbI_3$) perovskite films is studied by steady-state and time-resolved PL to investigate the detrapping rate of trapped charge carriers and to determine the density of traps (Wen et al. 2016). Taking into account the free electron–hole recombination, the Shockley–Read–Hall (SRH) recombination via trap states, and the Auger recombination, a model is established for the rate equations, which are used for fitting the experimental PL data (PL intensity versus the excitation intensity). In addition, TRPL spectra show two decay components: the fast component of decay time between 5 and 20 ns is attributed to defect trapping while the slow component of decay time between 150 and 230 ns is attributed to recombination process. A defect density of $\sim 7 \times 10^{16}\, cm^{-3}$ with a trapping time of $\sim 1.6 \times 10^{-6}\, s$ for free electrons and $\sim 6.7 \times 10^{-8}\, s$ for free holes is determined for the studied perovskite.

Other investigations of defects in $MAPbBr_3$ by the FL technique show the influence of the crystal size on the emission process (Droseros et al. 2018). Upon crystal reduction (from polycrystalline size $\sim 1\, \mu m$ to nanocrystalline size $\sim 10\, nm$), the photoluminescence quantum yield (PLQY) of the perovskite increases and the enhancement is explained by the reduction of the surface trapping and the increase of the photogenerated excitons.

The steady-state PL and TRPL measurements are also used to study physical processes, which are responsible for the changes in emission of perovskite materials. Enhancement of the emission in chlorine-doped $MAPbI_3$ is explained by the suppression of traps associated with iodine vacancies (Nan et al. 2020). Other studies report on an enhancement of the emission upon exposure of the perovskite to oxygen (Tian et al. 2015) or humidity (Brenes et al. 2017) and a decrease in the PL intensity of $MAPbI_3$ perovskite placed in low vacuum conditions ($< 10^{-4}$ mbar) (Fang et al. 2016). Several possible explanations for the observed changes in the PL intensity are given; however, it appears that the experimental conditions are important and may influence the measurement results. Therefore, definitive conclusions on the physical processes in studying the defects of perovskite by PL spectroscopy cannot be drawn without further complementary experiences with more defined measurement protocol.

Among the numerous investigations of perovskite materials, a review on the characterization of perovskites by the FL technique discusses the impact of the bimolecular recombination coefficient and the kinetic parameters, which allow for a better understanding of the operation of perovskite solar cells in a steady state (Kirchartz et al. 2020).

4.2.2. *Thermally stimulated luminescence (TSL) spectroscopy*

TSL is the emission of light from a semiconductor when it is heated. However, the light emission is not spontaneous as in the case where materials are heated to incandescence, but is produced after a previous absorption of energy from the radiation of a semiconductor and is then triggered by heating the material. The principle of the TSL measurement is schematically presented in Figure 4.4. The sample in an equilibrium state at room temperature is cooled down to a low temperature. Then, it is excited by radiation and electrons from the valence band are promoted to the conduction band by the generation process. Some charge carriers (electrons or holes) can be trapped in localized levels during this phase, and because of the low temperature of the sample, thermal emission from traps is negligible. Upon heating, the trapped charges – for instance, the electrons – are released. A released electron can recombine radiatively with holes through a recombination center. The TSL spectrum (also called the *glow-curve*) is the plot of the luminescence intensity as a function of temperature $I(T)$.

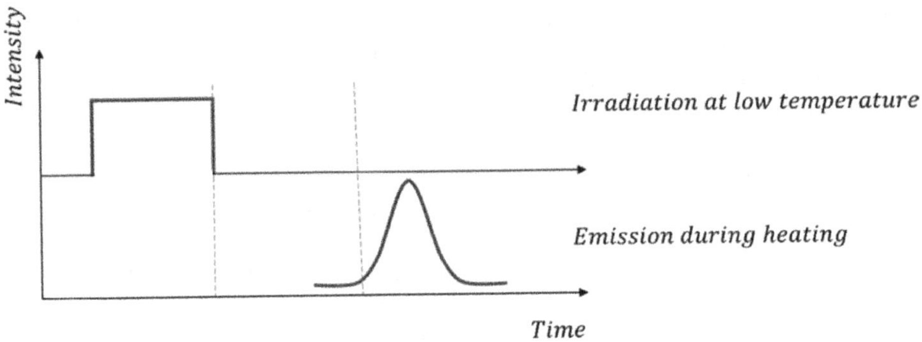

Figure 4.4. *Principle of the TSL measurement. For a color version of this figure, see www.iste.co.uk/nguyen/defects.zip*

4.2.2.1. *Equations describing the TSL curves: Randall–Wilkins first-order equation (Randall and Wilkins 1945)*

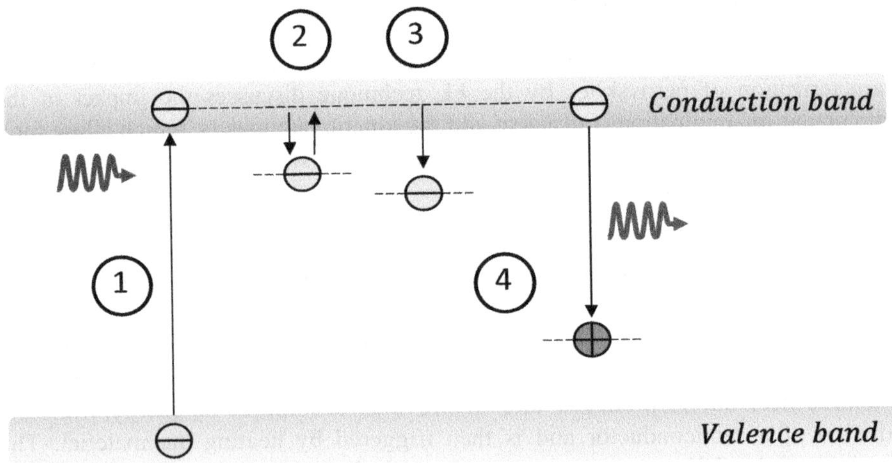

Figure 4.5. *TSL process: 1 – creation of excitons during irradiation; 2 and 3 – electron trapping by defect centers. In 2, release of trapped electrons from shallow traps; 4 – radiative recombination with emission of a photon from the defect center. For a color version of this figure, see www.iste.co.uk/nguyen/defects.zip*

The light emission in the TSL experiment results from radiative recombination of excitons created in the semiconductor during the excitation phase of the material by radiation of sufficient energy to promote an electron from the valence to the conduction band. In organic semiconductors, with the band gap being sufficiently small, the UV light is currently employed to excite the material. When there is a large number of defects, the photogenerated electrons in the conduction band, or holes in the valence band, can be captured by the defects in electron or hole traps. A trapped electron can be thermally raised to the conduction band and, while being free, move in the band and can recombine with a trapped hole. If the recombination process is radiative, then the luminescence will be observed due to the electron traps as shown in Figure 4.5.

Trapped charges at defect centers near to an allowed band can return to this band if sufficient energy is provided by thermal supply (case 2 in Figure 4.5) by a controlled process of heating. This can be done by a source with a constant or *linear heating rate* $\beta = \frac{dT}{dt}$ (of unity $K.s^{-1}$) providing a temperature at time t given by:

$$T(t) = T_0(1 + \beta t) \quad [4.10]$$

where T_0 is the initial temperature.

As the temperature increases, the charge carriers at various levels are gradually released from the trapping centers. The release probability can be described by the Boltzmann factor $\exp\left(-\frac{E}{kT}\right)$, where E is the activation energy of the process, which is the energy difference between the trap and the allowed band edge. For trapped electrons of density n, the release rate can be written as (see equation [3.32]):

$$\frac{dn}{dt} = -n \times \nu\, exp\left(-\frac{E}{kT}\right) \quad [4.11]$$

where ν is the *frequency factor* or *attempt-to-escape frequency*. The order of magnitude of this factor is in the range 10^{10}-10^{14} Hz.

This equation can be written as:

$$\frac{dn}{n} = -\nu\, exp\left(-\frac{E}{kT}\right) \times \frac{dt}{dT} dT = -\frac{\nu}{\beta}\, exp\left(-\frac{E}{kT}\right) dT \quad [4.12]$$

Solving this equation, we obtain:

$$n(t) = n_0 \exp\left[-\frac{\nu}{\beta} \int_{T_0}^{T(t)} exp\left(-\frac{E}{k\theta}\right) d\theta\right] \quad [4.13]$$

In this expression, θ is a dummy variable representing the temperature, which disappears upon integration.

The luminescence in this case is thermally stimulated and reflects the defects inside the semiconductor, and the light intensity is supposed to be proportional to the rate of electron release from the trapping centers. Therefore, we can write:

$$I_{TL}(t) = n_0\, \nu\, exp\left(-\frac{E}{kT}\right) \exp\left[-\frac{\nu}{\beta} \times \int_{T_0}^{T} exp\left(-\frac{E}{k\theta}\right) d\theta\right] \qquad [4.14]$$

where n_0 is the number of electrons initially trapped.

Equation [4.14] is called the Randall–Wilkins first-order equation and expresses the variation of the luminescence as a function of time that results from the radiative recombination due to defect states in the semiconductor. We can see that when the temperature increases, the luminescence intensity $I_{TL}(t)$ increases because more trapped charges are released, thus enhancing the recombination rate. The intensity $I_{TL}(t)$ will reach a maximum and decrease when most of the trapped charges are released. The TSL spectrum shows a luminescence peak at a characteristic temperature T_m of a trap level in the gap. The activation energy E can be determined from the initial rise of the thermo-luminescence curve from equation [4.14]. The conditions for applying equation [4.14] are the following: (1) the temperature interval ΔT used for the initial rise should be small and (2) the number of released carriers should be small as compared to the total number of trapped carriers. The activation energy is given by:

$$E = -\frac{d[ln(1/T)]}{d(1/kT)} \qquad [4.15]$$

The condition for the maximum intensity of the spectrum is found by the condition $\frac{dI}{dt} = 0$ which leads to the following expression:

$$\frac{\beta E}{kT_m^2} = \nu\, exp\left(-\frac{E}{kT_m}\right) \qquad [4.16]$$

From equation [4.16], we can write:

$$ln\left(\frac{T_m^2}{\beta}\right) = \frac{E}{kT_m} + ln\left(\frac{E}{\nu k}\right) \qquad [4.17]$$

Plotting $ln(T_m^2/\beta)$ against $1/kT_m$ should provide a straight line of slope E and an intercept with the y-axis at $E/\nu k$. The frequency factor at the maximum TSL peak can be determined from the experimental spectrum in this case.

The Randall–Wilkins model is based on a simplified hypothesis involving radiative combinations of trapped charges in a discrete trap level. Indeed, as we have seen, many factors and parameters can deeply modify the trapping process as well as the emission from these traps.

This model was improved with the consideration of the possibility for the released charge carriers to be retrapped by empty luminescence centers with the same probability as for its initial trapping.

4.2.2.2. *Garlick and Gibson model: second-order equation (Garlick and Gibson 1948)*

This model considers that the released carriers from traps can be retrapped by the available empty trapping centers. Consider a semiconductor containing a total concentration of electron traps N, and let n be the concentration of filled traps. The concentration of empty traps that can capture an electron is then $(N-n)$.

The probability for a released electron to recombine with an empty trap and not to be retrapped is:

$$p = \frac{n}{(N-n)+n} = \frac{n}{N}$$

The electron release rate given by equation [4.11] is modified by:

$$\frac{dn}{dt} = -n \times \nu\, exp\left(-\frac{E}{kT}\right) \times p = -\frac{n^2}{N} \times \nu\, exp\left(-\frac{E}{kT}\right) \qquad [4.18]$$

From equation [4.14], the luminescence intensity with retrapping of carriers becomes:

$$I_{TL}(t) = \frac{n_0^2\, \nu\, exp\left(-\frac{E}{kT}\right)}{N\left[1+\frac{n_0 \nu}{N\,\beta}\int_{T_0}^{T} exp\left(-\frac{E}{k\theta}\right)d\theta\right]^2} \qquad [4.19]$$

Compare this expression to equation [4.14], several different and similar characteristics of the luminescence spectrum can be noted:

– when all the traps are filled, the temperature at which the maximum the luminescence peak is reached is the same for both processes;

– the area under the luminescence spectrum is proportional to the number of electrons initially trapped n_0;

– the initial rise of the luminescence spectrum is different in case of carrier retrapping, by the power of the n_0 term ($n_0{}^2\nu/N$ and $n_0\nu$).

Figure 4.6 shows the comparison of a first-order kinetic and a second-order kinetic TSL spectrum, built from a simulation of equations [4.14] and [4.19]. Due to the retrapping process, the second-order TSL spectrum is enlarged with a higher intensity at the high temperature side.

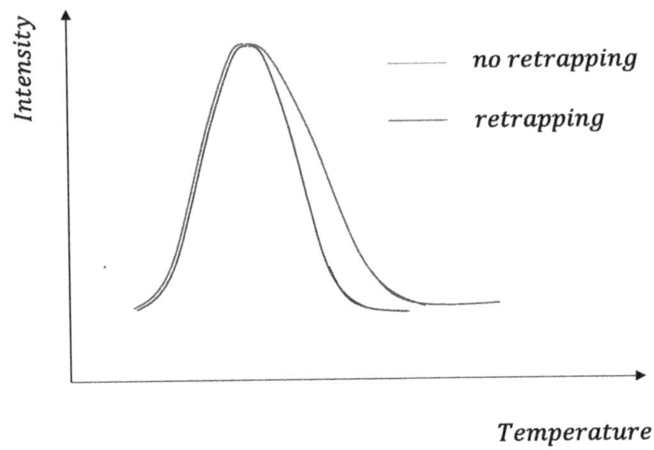

Figure 4.6. *Comparison of TSL spectra of first and second order (with and without retrapping of released charge carriers). For a color version of this figure, see www.iste.co.uk/nguyen/defects.zip*

In analyzing real systems, many complications may occur such as non-radiative recombinations, trap distributions, multitrapping, etc. which need more hypotheses, more parameters and more analyzed works. Please refer to the literature (McKeever 1981) for more details on these complex mechanisms. An advantage of the simplified models is that they can reliably describe the experimental TSL results with a reasonable number of parameters and allow for the estimation of the defects in the studied materials. In the case of the organic semiconductors, it would be necessary to consider the effect of the distribution of the localized states in the band gap on the trapping and detrapping processes.

4.2.2.3. *Trap parameters determined by the TSL*

In TSL experiments, the photogenerated carriers are obtained from the excitation of the material at low temperatures by a light source (usually an UV or mercury lamp with optical filters for light selection). The generated carriers fill the trapping centers, and the low temperature prevents them from escaping from the traps (negligible detrapping process). Then, the trapped carriers are gradually released by heating the system with a linear heating rate. The luminescence intensity due to radiative recombination of the released charges is recorded as a function time, which is related to the temperature in equation [3.40]: $T(t) = T_0(1 + \beta t)$. The experiment conditions are schematically shown in Figure 4.7.

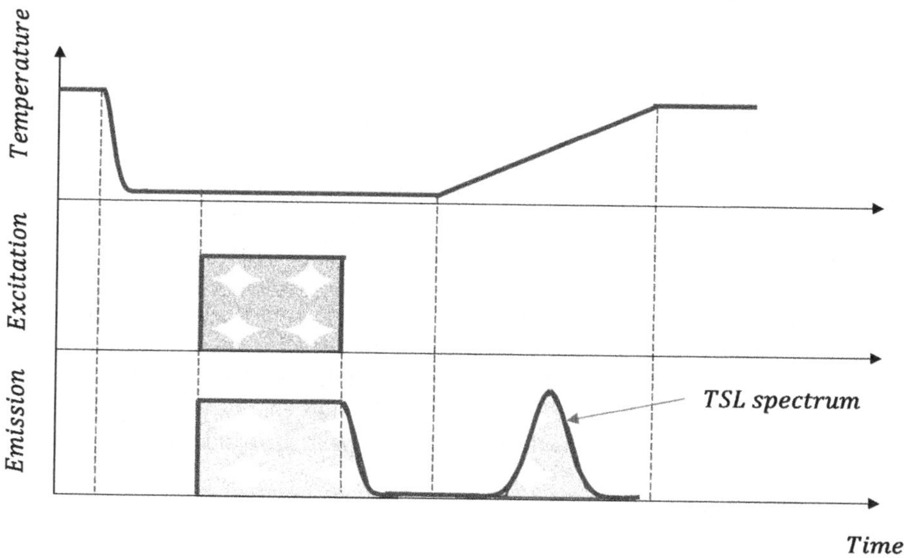

Figure 4.7. *Experimental parameters at different stages of the TSL spectroscopy. For a color version of this figure, see www.iste.co.uk/nguyen/defects.zip*

From the TSL spectrum, some trapping parameters can be determined by applying the simple models previously described.

According to the analysis of Garlick and Gibson, the initial rise of a TSL spectrum is dependent on the temperature following the expression:

$$I_{TL}(T) = C\ exp\left(-\frac{E}{kT}\right) \quad [4.20]$$

where C is a constant of all other parameters, which are supposed to be independent of the temperature.

A plot of ln(I) versus 1/T for the initial rise (or low temperature), part of the spectrum provides the activation energy or the trap depth $E = E_T$. It should be kept in mind that equation [4.20] is valid for a discrete trap hypothesis and the activation energy is determined for a TSL peak, which is well defined and separated from the other peaks. In real TSL spectra, because of the trap distributions in the band gap, several peaks are usually found and should be separated to allow for the determination of E_T. On the contrary, the trapped carriers' concentration is supposed to be constant independently of the temperature, which is not valid when the temperature increases over a certain limit (carrier detrapping). It is then necessary to separate the different components of the spectrum when applying the initial rise method to determine the trap depth E_T.

To do so, the fractional glow technique (Gobrecht and Hofmann 1966; Tales 1981) can be applied to resolve the TSL curve. The method consists of monitoring the TSL intensity during short heating and cooling cycles (for small temperature intervals of few degrees) in such a way that the carrier detrapping remains negligible, that is, the trapped carrier concentration n stays approximately constant. The small cycles are considered as temperature oscillations on a uniform heating ramp. The whole TSL is recorded and from the slope of the heating and cooling curves, the average depth of the trap states, which are emptied during each cycle, can be determined, and this make it possible to produce the energy distribution of the trapping centers in the band gap of the semiconductor. This technique is time consuming but provides a precise evaluation of the trap energy when traps are not well separated in the energy or the distribution of traps in the band gap is involved.

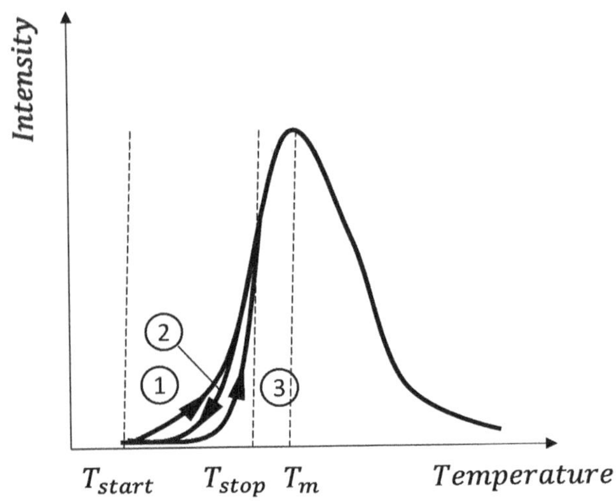

Figure 4.8. *$T_m - T_{stop}$ technique: heating cycle scheme of the sample: 1 – prerelease phase defining T_{stop}; 2 – cooling phase and 3 – main heating phase defining T_m*

Another simple way to investigate the composition of the TSL spectrum is called the $T_m - T_{stop}$ technique (McKeever 1981), which is also used for studying the trap profile in the band gap. Firstly, the sample is heated to an intermediate temperature called T_{stop}. It is then rapidly cooled and reheated until the whole glow curve is obtained. The temperature T_m of the maximum peak intensity is recorded. Next, the procedure is repeated with a slight increase of the T_{stop} temperature each time (Figure 4.8). For each cycle, a number of trapped charges are released and the intensity of the TSL decreases by gradual emptying. The temperature T_m does not change for first-order peaks because the activation energy does not change during the carrier detrapping; only their number changes. For the second-order peaks, a small shift in T_m to higher temperatures is observed with increasing T_{stop} and is explained by the gradual filling of traps upon successive cycles.

The plot of T_m as a function of T_{stop} allows for the identification of individual peaks in the TSL spectrum:

– for a single discrete trap level and in case of first order kinetics, the plot is simply a flat segment as T_m remains constant when T_{stop} increases;

– for traps with overlapping peaks, each individual component is represented by a flat segment (T_m is constant) and different individual peaks are represented by flat segments separated by vertical temperature intervals as shown in Figure 4.9 (the staircase shape);

– for a trap distribution, the flat segments are closely spaced resulting in a straight line of slope ~1.

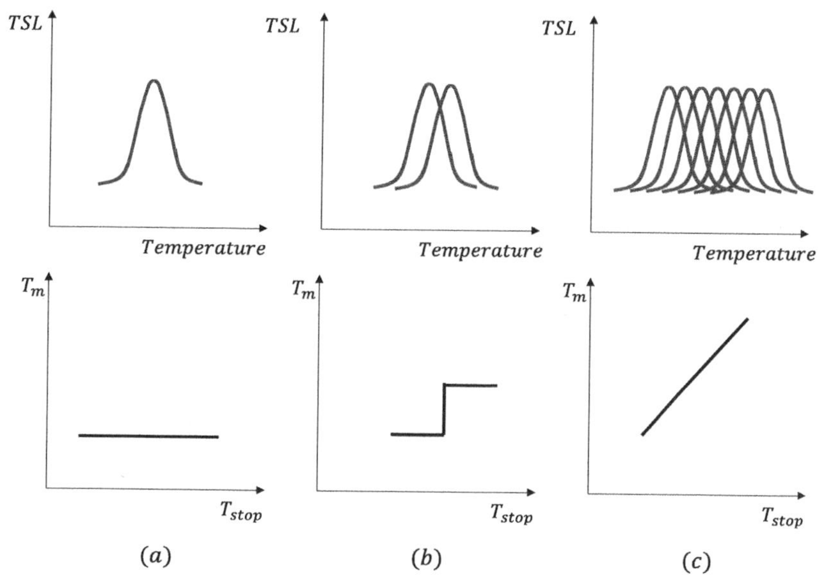

Figure 4.9. *TSL spectra and $T_m - T_{stop}$ plot: (a) single discrete trap TSL spectrum, (b) staircase-shape plot of TSL spectrum with overlapping peaks and (c) trap distribution TSL spectrum*

To illustrate the TSL spectra measured by this technique, a set of glow curves obtained in aluminosilicate glasses are shown in Figure 4.10. The first peak or inflection point T_m is recorded, and the experiment is repeated with a different T_{stop}: values are ranged from 293 to 653 K. From the initial slopes of the spectra, the activation energy E_t is estimated which increases when the T_{stop} increases since deep traps are gradually emptied. Fitting the curve using Gaussian functions, four distributions are determined with the activation energy E_t of 0.82, 1.08, 1.83 and 1.38 eV.

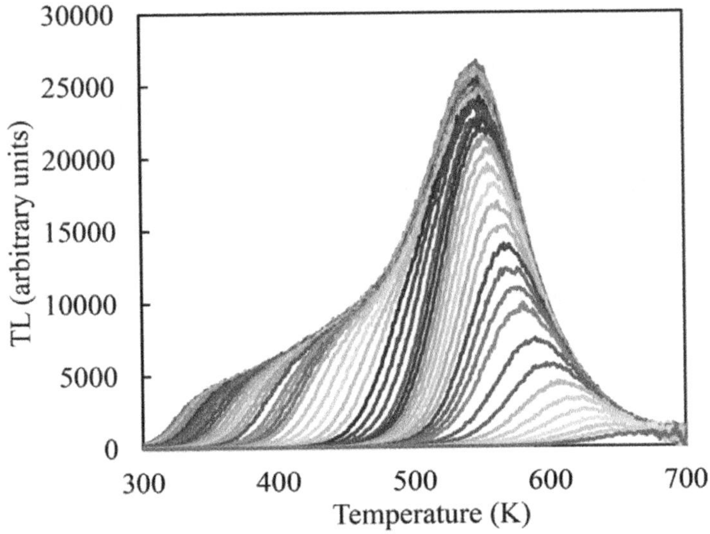

Figure 4.10. *$T_m - T_{stop}$ glow curves recorded aluminosilicate glass from 293 K to 748 K at a heating rate of $\beta = 2Ks^{-1}$ using different T_{stop} temperatures (from McKeever and Sholom 2021). For a color version of this figure, see www.iste.co.uk/nguyen/defects.zip*

4.2.2.4. *TLS spectrum fitting using Gaussian-shape TL peaks*

In disordered materials including organic semiconductors, the localized states are distributed in the band gap. Therefore, the trap levels are distributed over an energy range of $(E_1 - E_2)$. In this case, the luminescence intensity of equation [4.14] is modified to take into account the trap distribution.

We can write (McKeever and Sholom 2021):

$$I_{TL}(T) = \int_{E_1}^{E_2} g(E)\, I_{gw}(T) dE_T \qquad [4.21]$$

where $g(E)$ is the trap state distribution function, and $I_{gw}(T) = \nu\, exp\left(-\frac{E_t}{kT}\right) \exp\left[-\frac{\nu}{\beta} \times \int_{T_0}^{T} exp\left(-\frac{E}{k\theta}\right) d\theta\right]$.

For a TSL spectrum composed of separated glow curves, the TL intensity is given by:

$$I_{TL}(T) = \sum_{E_1}^{E_2} g(E)\, I_{gw}(T) \qquad [4.22]$$

A large number of experimental TL glow curves show Gaussian-shaped peaks for disordered semiconductors with high trap densities. Such a distribution can be understood as a consequence of the typical disorder in these materials, especially the organic ones. The $g(E)$ function can be therefore applied to describe the TSL spectrum of these materials. For a Gaussian trap-peak, the distribution function is:

$$g(E) = \sqrt{\frac{1}{2\pi}\frac{N_m}{\sigma}} \exp\left[-\frac{(E-E_m)^2}{2\sigma^2}\right] \qquad [4.23]$$

where E is the trap energy, σ is the width of the trap distribution, and $N_m$is the maximum concentration of traps of energy E_m. Superposition of Gaussian-like glow curves will result in a Gaussian distribution of defect states. However, it should be noted that the TLS spectrum is asymmetric according to the Randall–Wilkins analysis. Therefore, a fitted TSL spectrum by using equation [3.88] to represent the glow curves is generally approximate.

Although the Gaussian distribution of traps is commonly chosen for analyzing and interpreting defect measurement results in organic materials, other types such as uniform or exponential distributions are also possible, especially in those with a large width of trap distribution.

4.2.2.5. *TSL measurements of organic semiconductor and devices*

The TSL technique has been applied for determining the localized states in polymethyl-phenylsilylene (PMPSi) over a temperature range from 4.2 to 350 K, using a UV light source for the excitation of the material (Kadashchuk et al. 1998). The TSL spectrum is recorded by using the fractional technique and the mean activation energy E is obtained by using the initial rise method. The temperature dependence of the activation energy is linear, suggesting a distribution of localized levels in the band gap of the polymer. The experimental parameters of defects are determined: E = 0.19 eV and $S = 10^{10}\, s^{-1}$.

The bulk trap state properties of small-molecule materials NPB and Alq3 have been measured TSL method (Forsythe et al. 1998). The TSL spectra for both organic materials show significant trap distributions over the temperature range from 10 to

300 K. In NPB, four energy levels of trap states centered from 0.05 to 0.20 eV are retrieved and in Alq3, three TSL peaks are related to a distribution of trap states from 0.13 to 0.25 eV and a fourth trap level corresponding to an isolated trap state at 0.07 eV. For both materials, the trapping mechanism involves a combination of first- and second-order emission.

Small molecules and polymers PPQ and TPQ used as electron transport layers in OLEDs have also been studied by the TSL technique to determine the defect levels (Imperia et al. 2001). The TSL measurements are performed in the temperature range from 77 to 450 K on samples, which are excited by a Xenon lamp. Trap levels are found at E = 0.03, 0.11 and 0.18 eV for PPQ and E = 0.24 eV for TPQ. However, the nature of the trap type (electron or hole) and the trap densities could not be determined from the technique used.

Application of the fractional TSL measurements for determination of the electronic density of states (DOS) of organic materials has been performed on two series of carbazole-phenyl (CP) and carbazole-biphenyl (CBP) derivatives materials (Stankevych et al. 2021). Here, the TSL results are analyzed by the Gaussian disorder formalism, based on the thermal release of carriers trapped in levels located at the lower part of the DOS distribution. The DOS profile and the TSL spectra show similar trends and can be approximated by a Gaussian distribution. By comparing the TSL results obtained on a set of different organic materials, with the DOS parameters obtained from simulation, it is suggested that the DOS distribution in organic materials can be determined by the fractional TSL technique.

Despite their multiple advantages to obtain information on defects in materials, the TSL technique has some problems, which are due to the luminescence processes. Indeed, some materials are not luminescent and cannot be studied using the technique for investigating the defects. Furthermore, the light emission by the released charge from traps involves mainly those escaping from shallow traps, and the carriers trapped in deep traps would not be detected because they tend to recombine when they are captured. Therefore, the results obtained by TSL need to be cross-checked with complementary techniques. Finally, the TSL technique cannot be used for studying defects in multilayer structures such as organic electronic devices because of the presence of interfacial layers, which are not isolated to be analyzed. Such devices are more adapted to electrical characterization techniques as described below.

4.3. Electrical techniques

The capture and release of charge carriers affect the transport in the materials and can be detected by the variation of the charge transport characteristics under different aspects and investigation means. Generally, the determination of the defects is based on the measurement of the current intensity in the material, that is, the quantity of charge transported in the material per unit of time, and in several circumstances, on the measurement of the capacitance of the sample, that is, the quantity of charge stored in the material under an applied difference in potential. For the latter case, the material should have a capacitor configuration with a Schottky barrier, a blocking contact or a P–N junction.

Contrarily to the optical techniques, which do not need physical connections between the samples and the measuring apparatus, the electrical techniques require deposition of metal electrodes on the samples or devices to allow for applying an electric potential, which is provided by a power source. In other words, due to the electric contacts made on the sample, one or more interface region is created and can potentially constitute supplementary defect sources, which should be considered when performing the measurements and analyzing the results. Kinetic of charge carriers inside the material is also investigated for studying the defect states and thus studying the charge transport as a function of time is a useful technique to gain an insight into trapping processes.

The principal electrical trap measurement techniques include thermally stimulated current (TSC), space charge limited current (SCLC), impedance spectroscopy (IS) and deep-level transient spectroscopy (DLTS).

4.3.1. *Thermally stimulated current (TSC) technique*

The TSC technique is based on the release of the trapped charges in the band gap of a semiconductor by supplying a sufficient quantity of thermal energy which makes it possible for the carriers, electrons or holes, to escape from the traps and to be detected and measured.

The basic principle of the method is comparable with that of the TSL technique. However, in the TSC technique, the carrier trapping can be enabled either by electric excitation or by optical excitation, on the one hand, and the trap characteristics are determined by measurement of an electric current, on the other hand. As a result, several TSC measurement variations are built and applied for trap measurements in semiconductors (Chen and Kirsh 1981).

4.3.1.1. *Basic principle of TSC*

The TSC technique consists of measuring the current generated by trapped carriers on the return to the equilibrium state of a semiconductor which has been previously excited. There are four main steps for a typical experimental procedure (Figure 4.11):

1) at a high temperature T_H, the sample is polarized by application of a voltage V_B, for a time t_C sufficiently long to obtain a steady state for the current flow;

2) the sample is cooled down to a low temperature T_L (or T_0) while maintaining the applied voltage V_B until obtaining a steady state for the current flow;

3) the sample is short circuited at low temperature and kept at the temperature T_L until obtaining a steady state for the current flow;

4) the sample is heated with a constant heating rate β, and the current is recorded as a function of time until reaching the initial temperature T_H.

This technique was initially called the *ionic thermo-current* (ITC) (Bucci et al. 1966) for its application to crystals containing dipolar ionic defects.

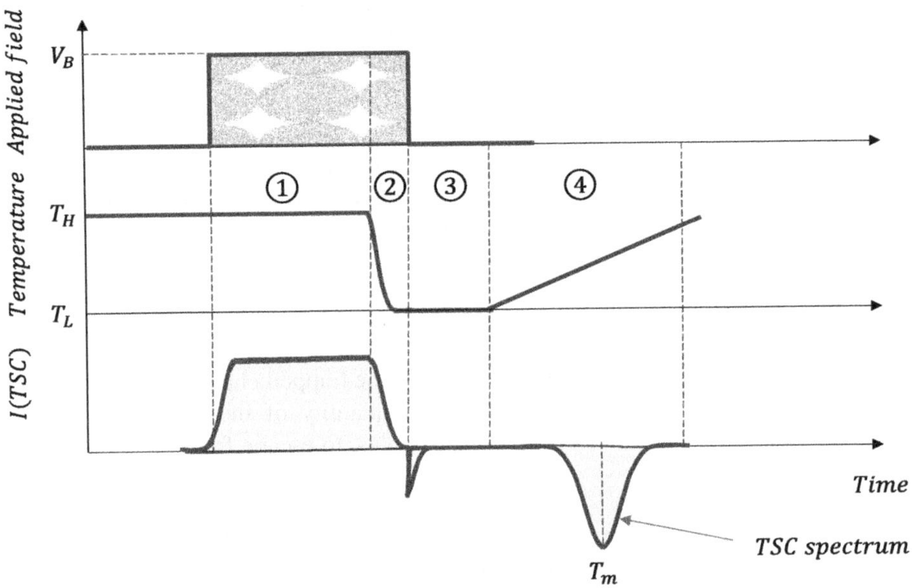

Figure 4.11. *Schematic representation of the principle of the TSC technique. For a color version of this figure, see www.iste.co.uk/nguyen/defects.zip*

Many other variations of the technique have been proposed and applied to study traps in semiconductors and amorphous materials. Among the most used, a technique similar to the TSL consists of creating a non-equilibrium carrier distribution by light (optical trap filling) in the sample kept at a low temperature. Then, thermal heating of the sample produces its relaxation to the equilibrium state and an excess current will be observed with an application of a small voltage to the sample (Driver and Wright 1963). A variation of this technique is called the *thermally stimulated polarization current* (TSPC) and is the inverse of TSC (McKeever and Hughes 1975). By using this technique, charge carriers are injected by an applied voltage into the sample kept in the dark (electrical trap filling). Firstly, the sample is cooled to a low temperature T_L. A voltage is then applied, and the sample is heated at a constant heating rate β. The current–temperature characteristic is recorded in which the current peak will appear due to the carrier detrapping.

The problem in exploiting the experimental results of these techniques is that the recorded current consists of two parts: (1) the detrapping current and (2) the conduction current, which may not be negligible when the sample temperature increases. Therefore, the recorded spectrum will not represent the released of trapped charge carriers in the sample (Argawal 1974). Thus, complementary data are needed to extract the detrapping current from the total current to determine the trap parameters.

The evaluation of the carrier trapping parameters from the TSC is analyzed in the following section. Fractional TSC measurements can be performed using the same procedure as in the TSL technique: the TSC spectrum is recorded during each temperature step (heating and cooling) and the activation energy is evaluated by the initial current rise to provide the energy distribution of traps inside the band gap.

4.3.1.2. *Evaluation of the trap parameters from the TSC spectrum*

4.3.1.2.1. Single trap level

Consider a semiconductor containing a single set of traps at energy E. For simplicity, the traps are supposed to be electron traps with a total density of N_t, and a capture cross-section of σ. In the TSC process, electrons are released in the conduction band of effective density conduction states N_C. For the concentration of trapped electrons n_t, the rate equations can be written as (Hearing and Adams 1960):

$$\frac{dn_t}{dt} = -n_t N_C \sigma v_{th}\, exp\left(-\frac{E}{kT}\right) + n_C(N_t - n_t) = -n_t \nu\, exp\left(-\frac{E}{kT}\right) + n_C(N_t - n_t) \quad [4.24]$$

$$\frac{dn_C}{dt} = -\frac{dn_t}{dt} - \frac{n_C}{\tau} \quad [4.25]$$

where $\nu = N_C \sigma v_{th}$ is the attempt-to-escape frequency, n_C is the free electron density, and τ is the lifetime of electrons in the conduction band.

Equation [4.24] describes the variation of the trapped electron density due to the thermal energy supply and to the retrapping process.

Equation [4.25] describes the variation of the electron concentration in the conduction band, which results from the electron detrapping from trap states and from the loss of electrons due to recombination with holes in the valence band, with respect to the carrier lifetime τ.

To solve the equation system, approximation is made on the overall rate of change of the electron concentration in equation [4.25] and is expressed by (Lewandowski and McKeever 1991):

$$\frac{dn_C}{dt} \ll \frac{n_C}{\tau} \rightarrow \frac{dn_t}{dt} \sim \frac{n_C}{\tau} \quad [4.26]$$

The electron density in the conduction can be written as:

$$n_C = \frac{n(T)(N_C \sigma v_{th})}{\frac{1}{\tau}+[N_T - n(T)]\sigma v_{th}} exp\left(-\frac{E}{kT}\right) \quad [4.27]$$

The $I_{TSC}(T)$ current intensity is expressed by:

$$I_{TSC}(T) = AF|e|\mu(T)n_C(T) \quad [4.28]$$

where A is the active area of the sample, F is the applied electric field, and μ is the mobility of the free electrons in the conduction band.

It is then possible to obtain the expression of the TSC current $I_{TSC}(T)$ for different hypotheses of retrapping, that is, for first- or second-order kinetics, and then use them to fit the experimental TSC spectra.

– For the slow retrapping process (first-order kinetics), the $I_{TSC}(T)$ current intensity is (Lewandowski and McKeever 1991):

$$I_{TSC}(T) = AF|e|\mu n_{t0}\tau v \, exp\left\{-\frac{E}{kT} - \exp\left[-\frac{v}{\beta} \times \int_{T_0}^{T} exp\left(-\frac{E}{k\theta}\right) d\theta \right]\right\} \quad [4.29]$$

where n_{t0} is the initial concentration of trapped electrons at the active level.

– For the retrapping process (second order kinetics), the $I_{TSC}(T)$ current intensity is (Garlick and Gibson 1948):

$$I_{TSC}(T) = \frac{AF|e|\,\mu n_0^2 \tau v \, exp\left(-\frac{E}{kT}\right)}{N\left[1+\frac{n_{t0} v}{N \beta}\int_{T_0}^{T} exp\left(-\frac{E}{k\theta}\right)d\theta\right]^2} \quad [4.30]$$

where n_{t0} is the initial concentration of trapped electrons at the active level. Equations [4.29] and [4.30] are established by assuming that the parameters μ, τ and v are temperature independent.

It can be noted that in both first- and second-order kinetic equations, the integral term is negligible in the low temperature region ($T \sim T_0 = T_L$). Therefore, in both cases, the TSC intensity can be approximated by $I_{TSC}(T) \propto \exp\left(-\frac{E}{kT}\right)$ independently of the kinetic order. Therefore, the activation energy of the trap levels can be determined by the initial rise method, as already explained in the TSL spectrum analysis.

The trap parameters determined by the TSC technique can be obtained by the following analysis:

– The trap density n_t is estimated from the total released charge Q measured by the area under the TSC spectrum:

$$\int_{T_0}^{T} I_{TSC}(T)dT \leq \Delta Q = |e|\, n_t\, (A \times d) \qquad [4.31]$$

where d is the thickness of the sample ($A \times d = V$ is the volume of the sample). Note that the estimated trap density n_t is only a lower limit of the actual density because all the traps are not filled, and the released carriers can recombine and they do not contribute to the measured TSC spectrum.

– The TSC peak maximum temperature T_m is determined by setting $\frac{dI_{TSC}(T)}{dT} = 0$. According to equation [4.17]: $\left(\frac{T_m^2}{\beta}\right) = \frac{E}{kT_m} + ln\left(\frac{E}{\nu k}\right)$, a plot of (T_m^2/β) versus $1/kT_m$ gives a straight line of slope E and an intercept with the y-axis at $E/\nu k$. Both the activation energy and the attempt-to-escape frequency can be determined by the temperature T_m.

Alternatively, the trap activation energy can be determined by the variation of the heating rate as proposed by Booth (1954). By differentiating equation [4.14] and setting the derivative equal to zero for the maximum temperature T_m, we can write:

$$exp\left(\frac{E}{kT_m}\right) = \frac{\nu k T_m^2}{\beta E} \qquad [4.32]$$

Using two heating rates β_1 and β_2 to measure the corresponding values T_{m1} and T_{m2} of the same sample, we obtain from equation [4.32]:

$$E = \frac{kT_{m1}T_{m2}}{T_{m1}-T_{m2}}\, ln\left(\frac{\beta_1 T_{m2}^2}{\beta_2 T_{m1}^2}\right) \qquad [4.33]$$

Once the activation energy is known, the attempt-to-escape frequency can be found by substituting E back into equation [4.32].

Another evaluation method (called the T^4 method) of the trap activation energy is derived from equation [4.32] by assuming that the trap capture cross-section σ and the carrier lifetime τ are constant at the maximum temperature (Bube 1960). The trap energy can be expressed as a function of T_m by:

$$E = kT_m ln\left(\frac{T_m^4}{\beta}\right) \quad [4.34]$$

Several models have been undertaken to exploit the TSC results for obtaining information of the trap parameters. They are more sophisticated than the fundamental models for treating the first- and second-order kinetics of traps. At the same time, more hypotheses and more variables have been introduced to deal with complex TSL and TSC spectra and to determine the basic trap parameters (activation energy, capture-cross section and trap density). The analysis methods of these parameters have been described by Nicholas and Woods (1964), who also compared the results obtained by various methods performed on the same material.

The experimental trap parameters for a given material are found to be strongly dependent on the conditions of preparation and the physical history of the samples. Therefore, these results should be cross-checked with others, which are obtained by complementary analysis techniques in order to be considered as reliable. To deal with complex trapping phenomena, a reasonable procedure would consist of identifying then separating the possible trapping processes in order to study them separately. In this way, we can apply basic models to treat the TSL and TSC spectra to determine reliably the trap parameters.

4.3.1.2.2. Distributed trap levels

To study materials having a trap distribution in the band gap, fractional methods can be used to resolve the TSC spectra. The principle of the method is the same used for the TSL, that is, to obtain the trap parameters (density and activation energy) by recording the TSC spectrum for a defined series of heating–cooling $(T_{start} - T_{stop})$ cycles over a chosen temperature range. The trap density is determined by measuring the number of released charges corresponding to the area under the TSC spectrum. The activation energy is determined by the initial rise method. Note that in the TSC technique, several configurations are possible for applying the voltage to electrically excite the sample. When performing the TSC measurements, the protocol for applying the voltage should be the same for each heating and cooling cycle (with or without light, with or without bias V_B).

Once the trap parameters are determined, the experimental distribution can be obtained by plotting the characteristic of the trap density against the activation energy $n_t(E)$. Then, the characteristic is fitted using a mathematical function $g(E)$. (Gaussian, exponential or uniform) as already explained for the TSL technique. It should be noted that for fitting the experimental data by any kind of mathematical function, there should be physical evidence of overlapping spectra (apparent TSC peaks) or trap-distribution shape spectrum as shown in Figure 4.9. In case of doubt, fractional TSC should be performed to identify the spectrum components. Without these observations, using mathematical functions for fitting the TSC curves remains purely a speculation and not facts.

Figure 4.12 shows the steps for fitting experimental TSC spectra using the fractional method. In this example, the energy distribution is assumed to be composed of two distributions and can be fitted by using Gaussian functions. The final spectra are exploited to yield the trap parameters (density and activation energy) of each distribution, from which the trapping process as well as the nature of the trapping species can be discussed or speculated.

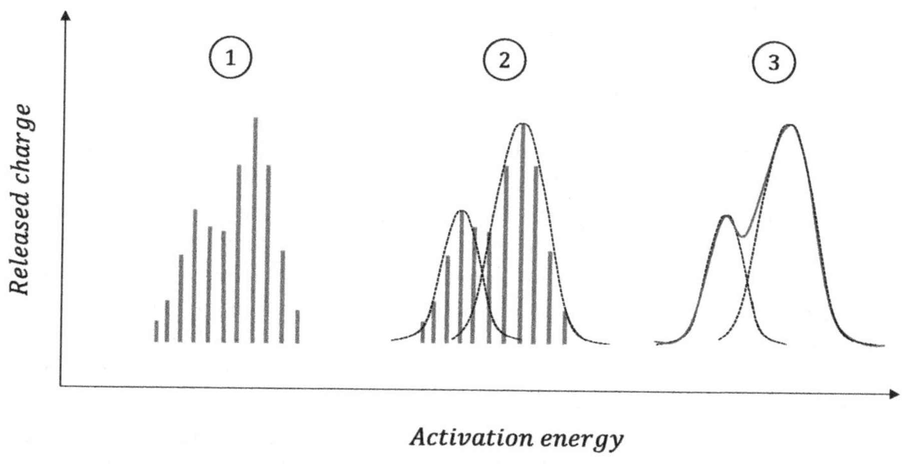

Figure 4.12. *Illustration example of the fitting steps of a TSC spectrum: (1) plot of the released charges versus the activation energy obtained from the fractional measurements, (2) fit of the data using two distribution functions and (3) decomposition of the energy distribution into two components, each of them can be related to a distinct trapping species or process. For a color version of this figure, see www.iste.co.uk/nguyen/defects.zip*

Figure 4.13 shows the fractional TSC spectra obtained in a polyfluorene-based diode and the Gaussian shape trap distribution using the fitting technique described above (Renaud et al. 2008).

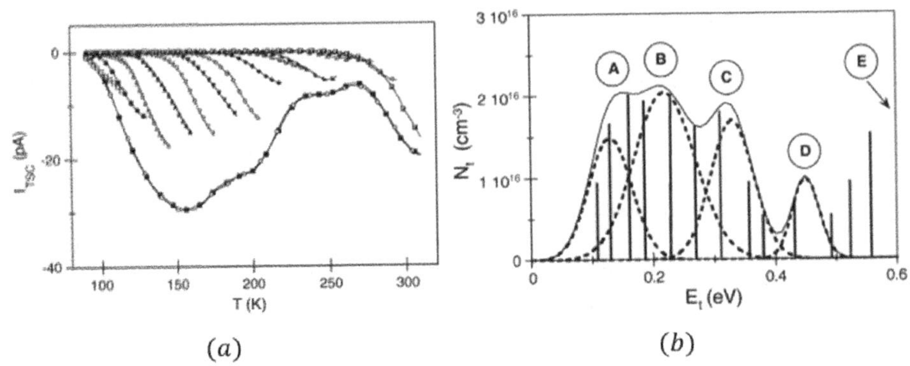

Figure 4.13. *(a) Fractional TSC spectra in a ITO/PEDOT/PF-N-Ph/Al diode in the temperature range 90–320 K with T_H = 300 K, V_C = + 6 V at different heating cycles, (b) Distribution of traps in a ITO/PEDOT/PF-N-Ph/Al diode obtained from fractional TSC measurements with V_C = + 6 V. Five Gaussian distributions are: A (E_t = 0.13 eV, N_t = 1.3 x 10^{15} cm^{-3}), B (E_t = 0.22 eV, N_t = 2.6 x 10^{15} cm^{-3}), C (E_t = 0.33 eV, N_t = 1.5 x 10^{15} cm^{-3}), D (E_t = 0.45 eV, N_t = 5.5 x 10^{14} cm^{-3}), and E (not resolved) (from Renaud et al. 2008, p. 7209)*

The TSC technique is efficient in studying defects in semiconductors, especially when using fractional measurements, information on the energy and the distribution of localized DOS can be obtained and provide useful data to investigate the transport process in materials, particularly in organic semiconductors and devices. As previously mentioned, to perform TSC measurements, the samples should be connected to an electrical circuit and metal electrodes have to be deposited onto the samples. The contact between a semiconductor and a metal is usually a supplementary source of defects, which should be considered in the determination of the trap parameters from the measurement results. It is possible to distinguish between the intrinsic defects of the materials and those created by and in the contact layers by methods that will be presented and described later on in this book. On the contrary, it is generally not possible to determine the nature of the defect centers (electron or hole traps) from simple TSC measurements. This is also valid for other electrical techniques used to study defects in organic semiconductors. We will see later that to address the problems, several solutions can be used. They are reliable and efficient but time consuming.

4.3.1.3. *TSC measurements of organic semiconductor and devices*

The TSC technique is intensively used to investigate defect states in organic semiconductors and devices. On the one hand, the technique is well adapted to electronic device structures, and on the other hand, experimental set-up is relatively simple to implement, and the results can be easily exploited. As examples for the investigation of defects using the TSC techniques, some typical organic materials and conjugated polymers are given below.

In small-molecule materials, trap measurements in Alq3, a frequently used electron transport and light-emitting material for small-molecule OLEDs are performed by the fractional TSC technique on $ITO/Alq3/Al$ diodes in a temperature range from -160°C to 10°C (Werner et al. 2001). The TSC measurement procedure is as follows: the sample is cooled to low temperature and a load voltage is applied. Then, an excitation by light of the sample is performed followed by an application of a read-out voltage to allow the current to cease. While keeping the applied read-out voltage, the sample is heated at a constant rate of 10 K/min and TSC spectra are recorded. The determined density of traps is $n_t = 1.7 \times 10^{17} cm^{-3}$. Traps in small molecules used as transport layer materials such as α-NPD and 1-NaphDATA are investigated by TSC with similar procedures and experimental conditions (Steiger et al. 2001). The trap parameters are determined by fitting the experimental spectra, assuming a retrapping of the released carriers. Two trap levels are found in both materials with activation energies in the range from 100 to 140 meV (α-NPD) and from 75 to 195 meV (1-NaphDATA). The trap densities are of the order of $n_t = 2.0 \times 10^{15} cm^{-3}$ for α-NPD and $1.0 \times 10^{16} cm^{-3}$ for 1-NaphDATA. Comparison of the trap parameters in devices using the studied materials in transport layers, before and after exposure to humidity, makes it possible to identify the formation of new traps in 1-NaphDATA, which are supposed to be responsible for the premature aging of the diodes.

Traps in pentacene-based Schottky diodes *Al/pentacene/Au* have been investigated by TSC as a function of bias voltage (Kuniyoshi et al. 2003). Two trap levels are found at 0.75 and 1.15 eV. While the density of the former is bias independent, the density of the latter depends strongly on the applied voltage and suggests that the related traps are originated from the Al–pentacene interface.

In conjugated polymers, the TSC technique has been used for determining trap parameters of poly(phenylene-vinylene) (PPV) in *metal/PPV/metal* diode structures in a temperature range from 100 K to 293 K (Nguyen et al. 1993a). A trap distribution is determined with three peaks corresponding to energy levels of 0.35, 0.68 and 0.82 eV. The TSC spectra are dissymmetric as regards the polarity of the applied voltage and suggest trapping effects at the polymer–metal interface. Similar results have been obtained by TSC measurements in PPP-based devices with trap activation energies of 0.36 and 0.47 eV in $Al/PPP/Al$ diodes and 0.44 eV in $Cr/PPP/Cr$ diodes (Ettaik et al. 1993). The trap densities are of the order of $10^{15} cm^{-3}$ for both types of devices.

Investigations of traps in MEH-PPV Schottky diodes have been carried out by TSC technique using both light and applied voltage excitations to fill the traps (Kazukaukas et al. 2002). Two trap activation energy distributions are found in the ranges 0.207–0.355 eV and 0.75–0.91 eV. The shallow traps are attributed to surface trap states, and the deep traps are distributed over the sample bulk.

P3HT:PCBM blend is intensively studied for use as an absorber material in bulk heterojunction solar cells. TSC measurements have been performed in degraded $ITO/PEDOT\!:PSS/P3HT\!:PCBM/Ca/Al$ and $ITO/P3HT\!:PCBM/Al$ diodes (Kawano and Adachi 2009) in the temperature range from 90 K to 423 K. Three trap levels of energy of 0.71, 0.81 and 0.91 eV are found and respectively attributed to (1) the PEDOT:PSS/P3HT:PCBM interface, (2) the P3HT:PCBM/cathode interface and (3) the ITO/P3HT:PCBM interface. The degradation of the solar cells is suggested to be due to the charge accumulation at the interfaces formed by the P3HT:PCBM layer with the electrodes.

Characterization of electronic trap states in perovskite solar cells based on methylammonium lead iodide (MAPI) has been performed using the TSC technique (Baumann et al. 2015). The structure of the diodes is $ITO/PEDOT\!:PSS\ (or\ CuSCN)/MAPI/PCBM/C60/Au$. Two shallow trap levels with low temperature peaks and two deep trap levels with high temperatures are found. One of the deep trap levels (with TSC peak at 163 K) is assigned to the orthorhombic–tetragonal phase transition. The second deep level (with TSC peak at 191 K) of activation energy of $\sim 500\ meV$ and a density of $\sim 1 \times 10^{15} cm^{-3}$ is assigned to defects the tetragonal phase.

4.3.2. *Current–voltage measurements: space charge-limited current (SCLC)*

In TSL and TSC techniques, the trap parameters are determined by measurement of the released charge carriers from trap centers by supplying a quantity of thermal energy. By measuring the current intensity in an organic diode under an applied voltage at room temperature, it is also possible to obtain information on the trap parameters of the organic semiconductor. The current–voltage characteristic of the organic diode in this case is analyzed by applying the SCLC theory, which is initially developed for dielectric materials.

4.3.2.1. *Space charge-limited current (SCLC)*

The space charge is defined as the positive or negative charge, which fills the semiconductor space under an applied field. The occurrence of a space charge takes place when the number of injected charges from an electrode into the material is more than the space can accept. Then, the excess charge will form a space charge at the vicinity of the injecting electrode, which will create an electric field to reduce the charge injection. Once the space charge is created, the current is controlled by the volume of the material and not by the injected carriers from the electrode. It is called SCLC. Such a current is observed in semiconductors having the following characteristics: (1) low mobility charge carriers, (2) high charge injection from the electrode (ohmic contact) and (3) strong transport unbalance.

It can be then understood that organic semiconductors, which contain a high concentration of defects and low carrier mobility, can exhibit space charge characteristic under an applied field whenever the charge injection rate is higher than the recombination rate, that is, when all the traps are filled. The applied voltage at which all the traps are filled is called *trap-filled limited voltage* V_{TFL}.

To establish the SCLC as a function of applied voltage, let us consider the transport processes of the semiconductor in different steps of the voltage application. We assume that the carriers are holes of density p and of mobility μ_p, the density of trapped holes is p_t.

Furthermore, the following assumptions are made:

– Under an applied voltage, the charge carriers are injected from the injecting electrode placed at $x = 0$, the counter electrode being at $x = d$.

– The density of free carriers is described by Boltzmann statistics:

$$p = N_V exp\left[-\frac{(E_{Fp}-E_V)}{kT}\right] \quad [4.35]$$

– The density of trapped carriers is described by the Fermi–Dirac distribution function:

$$f_p(E) = \frac{1}{1+\exp\left[\frac{(E_{Fp}-E)}{kT}\right]} \quad [4.36]$$

– The electric field inside the solid follows Poisson's equation:

$$\frac{dF(x)}{dx} = \frac{|e|}{\varepsilon_0\varepsilon_r}[p(x) + p_t(x)] \quad [4.37]$$

– The current density is given by:

$$J(x) = |e|\,\mu_p p(x)F(x) \quad [4.38]$$

– Diffusion current is negligible.

4.3.2.2. *Ohmic conduction*

The electrode contact is supposed to be ohmic, that is, capable of supplying charge carriers as many as needed to the semiconductor. At a low applied voltage V, the current density follows Ohm's law and can be written as:

$$J_O = |e|p\mu_p F = \frac{|e|p\mu_p}{d} V \qquad [4.39]$$

where p is the carrier density, μ_p is the carrier mobility, and d is the thickness of the semiconductor.

It should be noted that the contact between an electrode and an organic semiconductor is not generally ohmic and an energy barrier usually exists at the metal/semiconductor interface. The charge carriers have to overcome the barrier before entering the semiconductor. Thus, the current density is limited by the injection at the contact and not by the bulk of the semiconductor, at least for a range of applied voltages. Therefore, equation [4.39] is only valid if the injecting contact is ohmic, and the carrier mobility cannot be estimated without this condition.

4.3.2.3. *Space charge-limited conduction*

In a trap-free semiconductor (no intrinsic carrier concentration), the current density obeys to the Mott–Gurney equation (Mott and Gurney 1940):

$$J_{MG} = \frac{9}{8}\varepsilon_0\varepsilon_r\mu_p\frac{V^2}{d^3} \qquad [4.40]$$

where ε_0 is the permittivity of vacuum, and ε_r is the relative permittivity of the semiconductor.

The current–voltage characteristic exhibits a quadratic behavior with $J_{MG} \propto V^m$ and $m = 2$. In equation [4.40], the carrier mobility μ_p is assumed constant. It is often observed a departure from the Mott–Gurney law with $m > 2$, and these values are interpreted as a field-dependent mobility, which increases with the increasing applied field. According to the Frenkel effect, a strong electric field causes the trap depth to be reduced, which increases the proportion of free carriers. The space charge current density is then changed into (Murgatroyd 1970):

$$J_{MGF} = \frac{9}{8}\varepsilon_0\varepsilon_r\mu_p\frac{V^2}{d^3}\exp\left[\frac{0.891}{kT}\left(\frac{|e|^3V}{\pi\varepsilon_0\varepsilon_r d}\right)^{1/2}\right] \qquad [4.41]$$

The transition between the ohmic to the SCL regime takes place at the applied voltage V_Ω such that: $J_O = J_{MG}$. This condition leads to the equation:

$$V_\Omega = \frac{8}{9}\frac{|e|pd^2}{\varepsilon_0\varepsilon_r} \quad [4.42]$$

4.3.2.3.1. Single trap level

For a semiconductor having a single trap level with a density of trapped carrier of p_t and a density of free carriers of p_V, the Mott–Gurney equation is modified to take into account the number of trapped carriers and becomes:

$$J_{MG} = \frac{9}{8}\,\varepsilon_0\varepsilon_r\mu_p\,\theta\frac{V^2}{d^3} \quad [4.43]$$

where $\theta = \frac{p_V}{p_V + p_t}$ is the ratio of free carrier density to the total carrier density. In a trap-free semiconductor, $\theta = 1$ (equation [4.40]).

When the free carrier density increases by the injection of charge under an applied field, the quasi-Fermi level E_{Fn} moves above the trapping level as most of the traps are filled. After all the traps are filled, the injected carriers will be free to move, and the current density will increase the passing to the trap free conduction regime. The trap-filled limit voltage V_{TFL} is the transition voltage between the trap-filled J–V characteristic and the trap-free J–V characteristic. Indeed, the transition region between the trap-filled and the trap-free SCLC behavior is generally characterized by a power law $J \propto V^n, n > 2$, and the regime is named *the trap-filled limited* (region 3 in Figure 4.14a).

V_{TFL} can be determined by integration of Poisson's equation over the volume of the semiconductor.

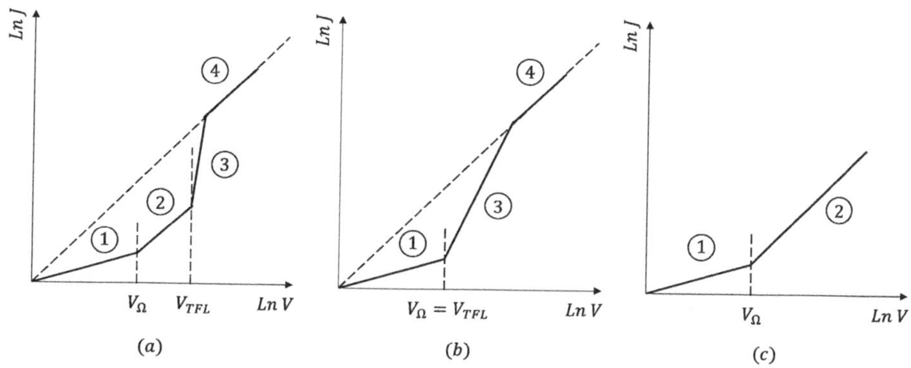

Figure 4.14. *Schematic plot of J–V characteristics showing conduction regimes in semiconductors exhibiting SCLC: (a) semiconductor containing shallow traps, (b) and (c) semiconductor containing deep traps with (1) ohmic regime, (2) trap-filled regime and (3) trap-filled limited regime and (d) trap free regime*

Consider a semiconductor containing hole traps of energy level E_t and total hole trap density P_t.

If the traps are shallow, then $E_t < E_{Fp}$, and according to equation [3.49], the density of unfilled hole traps is $P_T - p_t = P_T \exp\left(\frac{E_{Fp}-E_t}{kT}\right) \sim P_T$ if $E_t \sim E_{Fp}$. In this case:

$$V_{TFL} = \int_0^d F_{TFL} dx = \frac{|e| P_T d^2}{2\varepsilon_0 \varepsilon_r} \quad [4.44]$$

If the traps are deep traps, then $E_t > E_{Fp}$, and all injected charge carriers will fill the unoccupied traps. According to equation [3.49], the density of unfilled hole traps is $P_T - p_t = P_T \exp\left(\frac{E_{Fp}-E_t}{kT}\right)$. Therefore:

$$V_{TFL} = \frac{|e|(P_T - p_t) d^2}{2\varepsilon_0 \varepsilon_r} \quad [4.45]$$

In the case of deep traps, as all the traps should be filled by the injected carriers, the transition from the ohmic to the trap-free regime starts at the voltage V_Ω , that is, $V_{TFL} = V_\Omega$ (Figure 4.12b).

For organic devices, it can happen that all the traps could not be filled when using a range of voltages that can be applied and can withstand the semiconductors. In such cases, only the trap-filled regime will be observed in the J–V characteristic (Figure 4.12c).

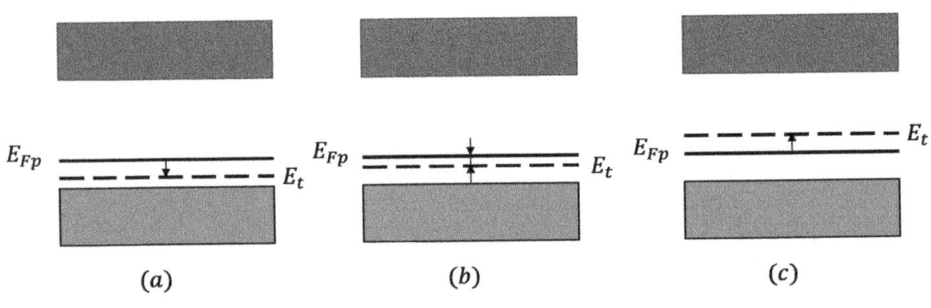

Figure 4.15. *Schematic representation of the energy band diagram of a semiconductor containing hole traps: (a) shallow traps* $E_t < E_{Fp}$, *(b) shallow traps* $E_t \sim E_{Fp}$, *and (c) deep traps* $E_t > E_{Fp}$. *For a color version of this figure, see www.iste.co.uk/nguyen/defects.zip*

Although it is possible to determine the trap parameters from the current–voltage characteristic analysis, a single trap level hypothesis does not reflect the actual energetic structure of organic semiconductors. Therefore, this result analysis gives only a rough estimation of the trap parameters and needs to be completed or checked with complementary experiments.

4.3.2.3.2. Exponential distribution of traps

In what follows, the term "distribution" means energetic distribution, even a spatial distribution $S(x)$ of the trap density should also be considered. This latter is not homogeneous because of the contact between the semiconductor and the electrodes, and their influence is all the more important when the semiconductor layer is thin. To take into account the spatial distribution, the thickness d of the layer should be replaced with an effective thickness d_{eff} in the different expressions of the current density. However, for simplicity, it is understood that the effect of the spatial distribution will be ignored, and we will use the true thickness of the semiconductor layer in all the equations.

On the contrary, it should be noted that in organic semiconductors, the transport is generally described with the hopping of free carriers between localized states in a Gaussian DOS. For the carrier trapping analysis, a trapped charge is captured and immobilized whereas a free charge is mobile and hopping between states in the DOS.

The exponential trap distribution can be described by the equation:

$$p(E,x) = \frac{P_T}{kT_C} exp\left(-\frac{E}{kT_C}\right) S(x) \qquad [4.46]$$

where T_C (or occasionally T_t) is a characteristic constant of the distribution, $S(x)$ is the spatial distribution of traps, and x is the distance from the injecting electrode. Alternatively, the *characteristic temperature* T_C can be replaced by the characteristic energy of the distribution $E = E_t = kT_C$, which represents the transport energy. The energy levels behind E_t act as hole trapping centers whereas the energy levels below E_t act as hole transport levels.

Integration of equation [4.46] between the limits E_{Fp} and ∞, the density of injected trapped holes is:

$$p_t(x) = P_T \left(\frac{p}{N_V}\right)^{T/T_C} S(x) \qquad [4.47]$$

The current density J can be then obtained by integration of equation [4.38] using the expression [4.46] for the trap distribution and with the boundary condition $V = \int_0^d F(x)dx$. We obtain:

$$J = |e|^{1-m}\mu_p N_v \left(\frac{2m+1}{m+1}\right)^{m+1} \left(\frac{m}{m+1}\frac{\varepsilon_0\varepsilon_r}{P_T}\right)^m \frac{V^{m+1}}{d^{2m+1}} = K\frac{V^{m+1}}{d^{2m+1}} \quad [4.48]$$

where the exponent $m = T_C/T$ is a temperature-dependent variable.

It can be seen that $ln\left(\frac{K}{d^{2m+1}}\right)$ is proportional to m and plotting $ln\left(\frac{K}{d^{2m+1}}\right)$ as a function of m will provide a straight line of slope S. If S is known, then the total trap density P_T in the semiconductor can be determined.

On the contrary, plotting m as a function of inverse of the temperature $1/T$ makes it possible to determine the transport energy E_t since $m = \frac{T_C}{T} = \frac{E_t}{kT}$.

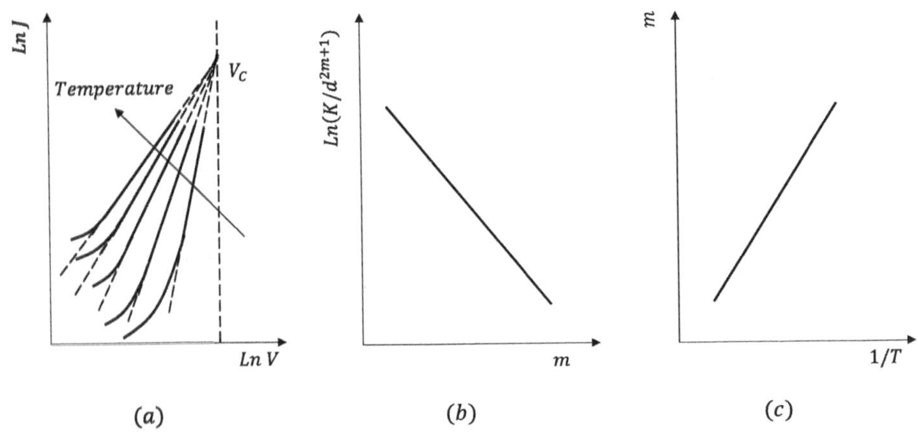

Figure 4.16. *Schematic plots of: (a) the SCLC current–voltage characteristics for different temperatures, (b) the variation of* $ln\left(\frac{K}{d^{2m+1}}\right)$ *with m, and (c) the variation of m with the inverse of the temperature*

Figure 4.16a shows schematically the plots of $ln\,(J)$ versus $ln\,(JV)$ for various temperatures from the slopes of which the values of m can be determined. Figure 4.16b shows schematically the plots of $Ln\left(\frac{K}{d^{2m+1}}\right)$ as a function of m from which the total trap density P_T can be determined. Figure 4.16c shows schematically the plot of m as a function of $1/T$ from which the transport energy E_t can be determined.

The total trap density P_T can also be determined by an analytical method (Kumar et al. 2003) by rearranging equation [4.48] as follows:

$$J = \left(\frac{\mu_p N_v |e| V}{d}\right) \times f(m) \exp\left[-\frac{E}{kT} ln\left(\frac{|e| P_T d^2}{2\varepsilon_0 \varepsilon_r V}\right)\right] \quad [4.49]$$

where the function $f(m)$ is defined by:

$$f(m) = \left(\frac{2m+1}{m+1}\right)^{m+1} \left(\frac{m}{m+1}\right)^{m} \frac{1}{2^m} \quad [4.50]$$

In practice, m is usually higher than 5 at room temperature, and the function $f(m)$ is approximatively constant (~0.5 for $m > 2$). The current density can be written by an Arrhenius equation with an activation energy E_a given by:

$$E_a = \frac{E_t}{k}\left[ln\left(\frac{|e| P_T d^2}{2\varepsilon_0 \varepsilon_r V}\right)\right] \quad [4.51]$$

It can be seen that when the applied voltage increases, the activation energy decreases since the trap states are gradually filled. At a critical or cross-over voltage $V = V_C$, all the traps are filled, and the activation energy becomes $E_a = 0$. From equation [4.51], we have:

$$V_C = \frac{|e| P_T d^2}{2\varepsilon_0 \varepsilon_r} \quad [4.52]$$

By plotting the current density–voltage characteristics of the diode at several temperatures, the cross-over voltage V_C can be determined by extrapolation and from the results, the total trap density can be calculated by equation [4.52].

4.3.2.3.3. Gaussian distribution of traps

The Gaussian trap distribution can be described by the equation:

$$p(E, x) = \frac{P_T}{(2\pi)^{1/2} \sigma_t} exp\left[-\frac{(E - E_{tm})^2}{2\sigma_t^2}\right] S(x) \quad [4.53]$$

where E_{tm} is the energy level corresponding to the maximum hole trap density (center of the Gaussian DOS), and σ_t is the standard deviation of the Gaussian function.

Here, the distinction between shallow and deep traps can be considered (Hwang and Kao 1976).

If the traps are shallow, then $E_{tm} < E_{Fp}$. The density of injected trapped holes is:

$$p_t(x) = \frac{p}{\theta_d} S(x) \qquad [4.54]$$

where $\theta_d = \frac{N_V}{P_T} exp\left[-\frac{E_{tm}}{kT} + \frac{1}{2}\left(\frac{\sigma_t}{kT}\right)^2\right]$.

The current density is obtained by integration of equation [4.38] using equation [4.54] with the boundary condition $V = \int_0^d F(x)dx$. We obtain:

$$J = \frac{9}{8}\varepsilon_0\varepsilon_r\mu_p\,\theta_d\frac{V^2}{d^3} = \frac{9}{8}\varepsilon_0\varepsilon_r\mu_p\frac{N_V}{P_T}\,exp\left[-\frac{E_{tm}}{kT} + \frac{1}{2}\left(\frac{\sigma_t}{kT}\right)^2\right]\frac{V^2}{d^3} \qquad [4.55]$$

Equation [4.55] is similar to equation [4.43] for the case of single trap level but with a different expression of the ratio of the free carriers to the total injected carriers θ_d.

If the traps are deep traps, then $E_{tm} > E_{Fp}$. The density of injected trapped holes is:

$$p_t(x) = \left[\frac{P_T}{2}\exp\left(\frac{E_{tm}}{mkT}\right)\right]\left(\frac{p}{N_V}\right)^{1/m} S(x) \qquad [4.56]$$

in which:

$$m = \left(1 + \frac{2\pi\sigma_t^2}{16k^2T^2}\right)^{1/2} \qquad [4.57]$$

Integration of equation [4.38] using equation [4.56] with the boundary condition $V = \int_0^d F(x)dx$ provides the SCLC current density:

$$J = |e|^{1-m}\mu_p N_v\left(\frac{2m+1}{m+1}\right)^{m+1}\left(\frac{m}{m+1}\frac{\varepsilon_0\varepsilon_r}{P_T}\right)^m\frac{V^{m+1}}{d^{2m+1}} = K_G\frac{V^{m+1}}{d^{2m+1}} \qquad [4.58]$$

Equation [4.58] has a similar form to that obtained for the exponential distribution of traps but with a different expression for the exponent m, given by equation [4.57].

From equations [4.55] and [4.58], it can be seen that the trap-limited transport in a semiconductor with a Gaussian distribution of traps can be approximated by either a single discrete trap level at low applied voltage or by an exponential trap distribution at high applied voltage.

The spatial dependence of filling of the Gaussian DOS can be considered through the distribution of the electric field, which is generally composed of the following three characteristic intervals (Arkhipov et al. 2001):

1) near the injecting contact, all the traps are filled, and the density of free carriers is higher that the density of trapped carriers. The current density corresponds to the trap-free SCLC;

2) for larger distances from the injecting contact, the density of charge carriers depends on the density of localized states below the quasi-Fermi level. These localized states act as deep traps for carriers as in the case of an exponential trap distribution;

3) far from the injecting contact, the density of charge carriers depends on the density of localized states above the quasi-Fermi level, which act as shallow traps. The current density corresponds to that of the SCLC with shallow traps.

The above analyses of the current–voltage characteristics lead to the determination of trap parameters with a few approximations and simplified hypotheses, and therefore, with a limited range of applied voltages for the validity of the analysis. A different approach for analyzing the SCLC characteristics called *the differential technique* is proposed (Nespurek and Sworakowski 1977) taking into account the position of the quasi-Fermi level in the solid and its variation during the measurements.

The electric field at the counter electrode ($x = d$) can be expressed by:

$$F(d) = \kappa_1 \frac{V}{d} \quad [4.59]$$

where κ_1 is a parameter representing the distance between the injected charge and the counter electrode with $1 \leq \kappa_1 \leq 2$.

The quasi-Fermi level depends on the charge injection and can be written as (Nespurek and Sworakowski 1980):

$$E_{Fp}(d) - E_V = kTln\left(\frac{d}{\kappa_1 N_V |e| \mu}\right) + kTln\left(\frac{J}{V}\right) \quad [4.60]$$

As a result, any change in the concentration of trapped carriers will modify the position of the quasi-Fermi level and make it possible to determine the energetic distribution of traps at the counter electrode. By changing the applied voltage, the new injected carriers are expected to be trapped, and the occupancy of the trap states is changed leading to a shift of the quasi-Fermi level towards the band edge. The SCL current–voltage characteristic can be analyzed to provide information on the DOS.

Introducing the slope of the SCLC characteristic defined by:

$$m = \frac{d(lnJ)}{d(lnV)} \quad [4.61]$$

The DOS distribution function $\frac{dp_t}{dE}$ can be obtained by differentiating the ratio θ of free carrier density to the total carrier density with respect to E_F (Nespurek et al. 2008):

$$\frac{dp_t}{dE} = \frac{\varepsilon_0 \varepsilon_r}{2|e|d^2 kT} \kappa_1 \kappa_2 \frac{V}{m-1} \quad [4.62]$$

where κ_2 is the ratio of the carrier concentration at the counter electrode to the injected carrier concentration with $\frac{1}{2} \leq \kappa_2 \leq 1$, and the $\kappa_1\kappa_2$ product is assumed to be in unity.

The activation energy of the trap centers can be determined by performing the $J(V)$ measurements as a function of temperature and by plotting the slope m as a function of inverse of the temperature. The density of traps is estimated from the DOS distribution function given by equation [4.62].

Analysis of defects in organic semiconductors by the current density–voltage –temperature characteristic measurements is possible, provided that the approximations made on the materials and devices are acceptable. In particular, the assumption that diffusion currents are negligible may not be justified and can lead to problematic situations by over- or underestimating the slope of the current density–voltage curves (Kirchartz 2013), and therefore, leading to incorrect results of trap parameter determination (Fisher et al. 2014). The voltage at which the diffusion can be neglected depends on several parameters such as the temperature, the trap density and the built-in voltage (Dacuna and Salleo 2011) which should be examined for the sample studied in order to achieve an accurate trap characterization by this method. Furthermore, as the current density–voltage characteristics are measured in bipolar devices using different contact layers (metal or organic) with the semiconductor, an internal electric field is created even in the absence of an applied voltage. This electric field may not be negligible and relates to a built-in potential (Mantri et al. 2013), which should be considered in the analysis of the experimental J–V characteristics.

4.3.2.4. *SCLC measurements of the organic semiconductor and devices*

The SCLC method is widely used for studying trap parameters in organic semiconductors and devices as it has been successfully applied for dielectric and amorphous materials. Compared to other characterization methods, the experimental set-up and procedure for the SCLC measurements are simpler. Indeed, for a simple diode structure (anode/semiconductor/cathode), the required equipment may consist only of a source measure unit for obtaining the SCLC characteristics and the results can be directly exploited. However, as we have seen previously, an accurate determination of the trap parameters requires a strict verification of the necessary conditions for applying expressions and equations to quantitatively obtain the trap parameters from the measurements. This is a major difficulty of using the SCLC method and explains why the assumptions are frequently made in the analyses and exploitation of the current density–voltage characteristics of devices. As a result, the results obtained from the SCLC methods are usually considered as approximated and need to be checked or completed by other measurement techniques.

There are numerous trap studies of organic semiconductors using the SCLC method, and we show hereafter some of the results obtained from the most typical materials used in organic devices.

For small molecules, trap characterization was investigated by the SCLC method in Alq3, which is the active organic material in $Al/LiF/Alq3/LiF/Al$ structures, in a temperature range from 93 to 333 K (Mizuo et al. 2002). By using the differential technique, the authors determined the trap density of Alq3 according to the trap distribution model used: (i) for an exponential distribution, the density of states at the LUMO energy is $2 \times 10^{21} cm^{-3}$ and (ii) for a Gaussian distribution, the maximum of the distribution is located at 0.26 eV below the LUMO level with a trap density of $1 \times 10^{18} cm^{-3}$.

Investigations of traps in α-NPD by SCLC were performed using hole-only devices of structure $PEDOT:PSS/(\alpha - NPD)/MoO_3$ and electron-only devices of structure $Al/(\alpha - NPD)/BaAl$ in a temperature range from 213 to 293 K (Rohloff et al. 2017). By simulation using Gaussian distribution traps, the authors determined trap parameters in electron-only devices with a trap density of $1.3 \times 10^{24} m^{-3}$, an activation energy of 0.67 eV and a distribution width of 0.2 eV. The hole transport shows a trap-free characteristic with a mobility of $2.3 \times 10^{-8}\, m^2V^{-1}s^{-1}$. The electron traps of high density are not structural defects but are believed to be extrinsic defects.

Mixed organic layers (MOL) composed of Alq3, α-NPD and rubrene can potentially be used in OLEDs for improving their performance and can be deposited by solution processing using chloroform as a solvent (Wang et al. 2011). The SCLC measurements in devices of the structure $Au/MoO_3/MOL/MoO_3/Au$ were performed in a temperature range from 233 to 273 K, and the results were analyzed by the differential method. For the high-applied field, the SCLC characteristics show a power law variation of the current density $J = KV^m$ and is assigned to a hole transport process with carrier trapping in the organic layer. The density of traps is estimated at $\sim 10^{18} cm^{-3}$ with a trap depth in the range of 0.55–0.65 eV from the HOMO level.

Investigations of defects in pentacene (Pc) films deposited by evaporation were carried out by the SCLC method applied to devices of structure $Al/Pc/Au$ and $Au/Pc/Au$ in the temperature range from 130 to 380 K (Lee and Gan 1977). To check the thickness dependence of the current density–voltage characteristics, pentacene films of thicknesses from 0.30 to 1.30 μm were used. The trap density ranging from 0.3 to $3 \times 10^{17} cm^{-3}$ and the trap depth from 0.96 to 1.02 eV were determined. These defects were assigned to the molecular disorder in the organic films.

For conjugated polymers, OLEDs using PPV as an active layer were characterized by current–voltage, impedance and transient conductance measurements (Campbell et al. 1997). The current–voltage characteristics of devices of structure $ITO/PPV/Al$ were measured from 30 to 290 K, and the results were analyzed by applying an exponential trap distribution model. A trap density of $\sim 5 \times 10^{17} cm^{-3}$ with a characteristic energy E_t of 0.15 eV is obtained and is in agreement with results obtained by transient conductance measurements. The authors suggested that the trap distribution could also be the tail of a Gaussian density of states through which the charge carriers hopping occurs and concluded that an exponential and Gaussian distribution is not distinguishable in the SCLC characterization.

A series of PPV derivatives including MEH-PPV, OC_1C_{10}-PPV and NRS-PPV were studied by the SCLC technique in devices of structure $Al/PPV\ derivative/Ba/Al$ (Nicolai et al. 2011). Analysis of the I(V) characteristics of the samples shows that the transport in the studied PPV derivatives can be *globally* described by a Gaussian trap distribution with a total density of states of $\sim 1 \times 10^{23}\ m^{-3}$, a trap depth located at 0.7–0.8 eV below the LUMO level and a Gaussian distribution width of 0.1 eV. Although these results are comparable to those obtained in PPV, it should not be generalized to other conjugated polymers because their structure is different, and the structural disorder and the related defects are inevitably different.

P3HT is well known as a promising material for applications in organic solar cells and field-effect transistors. Defects in diodes of structure $ITO/P3HT/Al$ were investigated by TSC and SCLC measurements in the temperature range from 80 to 300 K (Nikitendo et al. 2003). Based on the results obtained by the TSC method, two Gaussian distribution models were considered forming a total DOS energy distribution given by:

$$g(E) = \sqrt{\frac{2}{\pi}} \frac{(N-N_t)}{\sigma} exp\left[-\frac{E^2}{2\sigma^2}\right] + \frac{N_t}{\sqrt{2\pi}\sigma_t} exp\left[-\frac{(E-E_t)^2}{2\sigma_t^2}\right] \quad [4.63]$$

where E_t is the average trap energy, σ_t is the width of the trap distribution, N is the total density of states, and N_t is the total density of traps. The trap density determined in the P3HT samples is $N_t = 1.5 \times 10^{16} cm^{-3}$, the trap depth is $E_t = 0.55\ eV$, and the width of the distribution is $\sigma_t = 0.08\ eV$.

Blends of P3HT:PCBM were studied by the SCLC method to determine the trap parameters as compared to those of the polymer (Rizvi et al. 2014). The device structures used are: $ITO/PEDOT{:}PSS/P3HT{:}PCBM/Al$ (devices A), $ITO/ZnO/P3HT{:}PCBM/MoO3/Ag$ (devices B) and $ITO/P3HT/Al$ (devices C). For devices using P3HT, the trap parameters ($N_t = 4.4 \times 10^{16} cm^{-3}$, $E_t = 0.38\ eV$) are comparable to those obtained in Kuniyoshi et al. (2003). For devices using the P3HT:PCBM blends, the trap parameters are $N_t = 1.4 \times 10^{17} cm^{-3}$, $E_t = 0.37\ eV$ (devices B) and $N_t = 3.2 \times 10^{16} cm^{-3}$, $E_t = 0.53\ eV$ (devices C). Here, no conclusion could be made on the influence of the incorporation of PCBM to the polymer on the defect formation process. It was also indicated that the quality of the blend could not be controlled given the various values of the trap parameters for the same type of samples tested.

SCLC measurements in metal halide perovskite-based devices are usually performed to probe the defect densities and the carrier mobilities in the hybrid materials. The difficulty of using this technique for the perovskites lies in the fact that both electronic and ionic conductions take place during the measurement of the J–V characteristic, and the influence of mobile ions on the current density can lead to inaccurate results. Indeed, it is shown that mobile ions play a key role in the hysteresis of perovskite-based devices by piling up near the interfaces and partly shielding the bulk material from the applied voltage. Since the ionic movement is expected to be slower than that of the electronic carriers, it is possible to separate the flow of the species during the measurement by using a voltage pulse of a chosen width. The choice of the pulse duration can limit the ion movement and allows us to correctly measure the electronic current density. The pulse voltage SCLC measurements were applied to study methylammonium bromide $MAPbBr_3$ perovskite single crystals of structure $Au/perovskite/Au$ (Duijnstee et al. 2020). The J–V characteristics were recorded using pulse voltages of a width of 20 ms. The samples were short circuited for 7 minutes between two consecutive voltage sweeps to allow for the ion redistribution within the material. The J–V characteristics vary with the thickness of the samples indicating the SCLC conduction, but only the ohmic and trap-filled limited regions could be observed (similar to the regions 1 and 3 of Figure 4.14b). From the J–V characteristics, the trap density is determined as $2.8 \times 10^{12} cm^{-3}$, which corresponds to the lower limit of the trap density in $MAPbBr_3$.

Other SCLC measurements on $MAPbBr_3$ and $MAPbI_3$ perovskite crystals (Shi et al. 2015) reported much lower densities of traps respectively $5.8 \times 10^{9} cm^{-3}$ and $3.3 \times 10^{10} cm^{-3}$. Here, the devices have a typical structure of

Ag/Au/MoO³/perovskite/MoO³/Au/Ag, and the trap density was determined from the V_{TFL} voltage of the J–V characteristics. The $MAPbBr_3$ perovskite was also studied by the differential method (Pospisil et al. 2019), and three trap states were determined with activation energies of $0.63, 0.55$ and $0.38\,eV$ and with concentrations of $2.4 \times 10^9, 4.5 \times 10^9$ and $6.2 \times 10^{10}\ cm^{-3}$.

There is a strong discrepancy of the trap density determined by different studies, and other investigations are needed to gain insight into the effect of the material preparation and the analysis of the SCLC method, which is still not a very accurate technique to study defects in semiconductors.

Trap densities in perovskite films are significantly different from those in perovskite crystals. For MAPI solution-processed film, used as an active layer in the $ITO/PTAA/MAPI/Au$ structure, the trap density determined from the analysis is $\sim 1.0 \times 10^{16} cm^{-3}$ (Petrovic et al. 2019). Here, PTAA is used as a hole transport layer (HTL) for improving the stability of the diodes. The trap density was found to be slightly increased when the diodes were degraded by exposure to UV light. Investigations of defects in formamidinium lead triiodide ($FAPbI_3$) by the SCLC methods in the diodes of the structure FTO/$FAPbI_3$/Au (Son et al. 2018) reported a density of $\sim 1.0 \times 10^{16} cm^{-3}$. Compared to single crystals, perovskite thin films show defect densities of five to six orders of a magnitude higher that those determined in crystals. This difference is generally explained by the presence of grain boundaries, which act as charge carriers trapping centers. Large grain sizes usually relate to low trap densities and high charge carrier mobilities in perovskites films (Yang et al. 2018) and can be controlled by choosing appropriate materials used for electron transport layer (ETL) in perovskite-based devices.

4.3.3. *Impedance spectroscopy (IS)*

According to the analysis of the current–voltage SCLC characteristic of two-port diodes, the trap parameters of the semiconductor used as an active layer can be determined, provided that appropriate assumptions are made. By this technique, the carrier mobility can also be measured from the Mott–Gurney law (equation [4.40]) and in organic semiconductors, from the modified expression of the current density (equation [4.41]) with a field-dependent mobility. The measurement of the carrier mobility is essential for understanding the physical processes of organic semiconductors and devices, and allows the characteristics and performance of electronic devices to be improved. Given that the SCLC method requires rather strict experimental conditions for obtaining accurate results of both trapping parameters and carrier mobility, an alternative transport characterization method, impedance or admittance spectroscopy, is used to complete the SCLC technique through the analysis of the dynamics of charge carriers under the application of an alternative signal to the device.

4.3.3.1. *Basic principles of the impedance spectroscopy technique*

The method consists of measuring the capacitance of a diode, which is composed of a semiconductor sandwiched between two electrodes as a function of several parameters such as the temperature T, the direct applied voltage V_{DC} and the frequency f.

The capacitance exists in such devices by the formation of a *depletion region* with band bending formed in a P–N or a Schottky junction. The width W of this region and thus the capacitance C depend on the DC applied voltage. Under an applied voltage of V_{DC} (positive for forward bias), the junction capacitance is proportional to the inverse of the space charge width and is given by:

$$C_J = A\left[\frac{|e|\varepsilon_0\varepsilon_r N}{2(V_{bi}-V_{DC})}\right]^{1/2} \qquad [4.64]$$

where A is the surface of the sample, N is the doping concentration, which is supposed to be uniform in the semiconductor, and V_{bi} is the built-in voltage of the junction.

In the measurements by the impedance technique, a small AC voltage $v_{AC}(t) = v_0 \cos \omega t$ of amplitude v_0 and of frequency $f = \omega/2\pi$ is added to the steady-state voltage, and the complex impedance of the device is measured by the AC current i_{AC} in the sample. The applied voltage is considered as a small signal, which would not consequently disturb the thermal equilibrium of the device.

Let V_{DC} be the direct voltage applied to the sample, the total voltage is:

$$V(t) = V_{DC} + v_{AC}(t) = V_{DC} + v_0 \cos \omega t \qquad [4.65]$$

The total current in the sample is the sum of the DC and the AC currents and can be written as:

$$I(t) = I_{DC} + i_{AC}(t) = I_{DC} + i_0 \cos(\omega t + \varphi) \qquad [4.66]$$

where φ is the phase shift between the voltage and the current in the sample.

The complex impedance of the sample in the complex notation is:

$$Z(\omega) = R(\omega) + i.X(\omega) = |Z(\omega)|\exp(i\varphi) \qquad [4.67]$$

where $R(\omega)$ and $X(\omega)$ are the real and the imaginary parts (also called the *resistance* and the *reactance*) of the impedance respectively, $i = \sqrt{-1}$, and $|Z(\omega)|$ is the modulus of the impedance:

$$|Z(\omega)| = \sqrt{R(\omega)^2 + X(\omega)^2} \quad [4.68]$$

The *complex admittance* of the device can be written as:

$$Y(\omega) = \frac{1}{Z(\omega)} = \frac{i_{AC}}{v_{AC}} = \frac{R(\omega)}{|Z(\omega)|^2} - i\frac{X(\omega)}{|Z(\omega)|^2} = G(\omega) + i\omega C(\omega) \quad [4.69]$$

where $Z(\omega)$ is the impedance, $G(\omega)$ is the *conductance*, and $C(\omega)$ is the capacitance of the device. $\omega C(\omega)$ is also called the *susceptance.*

By analyzing and exploiting the real ($G(\omega)$) and the imaginary ($C(\omega)$) parts of the complex admittance of the device, it is possible to extract the information on the conduction process of the charge carriers and their kinetics in the devices. As a result, defect states can be studied from the admittance measurements of the devices.

4.3.3.2. *Impedance measurements of devices*

The impedance of the devices is usually measured by using a capacitance bridge or an impedance analyzer, which can generate signals of frequency in the range from millihertz (mHz) to megahertz (MHz) or more. The experimental set-up also includes a DC bias (usually incorporated to the bridge or analyzer), and measurements of the impedance as a function of temperature are usually useful.

Generally speaking, the presence of defects in devices can be observed by a step in the low frequency region of the capacitance spectrum, which is shifted to higher frequencies with increasing temperatures. This capacitance step results from an increase in charge due to the release of carriers from trapping centers with applied DC and AC voltages.

To detect and characterize defects in devices, the measurement of the impedance, and especially the capacitance, should be carried out in an appropriate range of frequency. Indeed, the trapped carriers are released from the trapping centers with a characteristic thermal emission rate (e_n or e_p in equation [3.39]), corresponding to a relaxation time (τ_n or τ_p). If the frequency of the applied AC signals is low ($f < 1/\tau_n(\tau_p)$), the charge carriers are released from the traps and contribute to the measured capacitance. In contrast, if the applied frequency is high, the emission rate will be masked because it cannot follow the rate of alternating voltage and will not contribute to the measured capacitance. The change between the two processes defines a *characteristic angular frequency* ω_0, which is related to the thermal emission rate of the charge carriers. In summary, for the frequency $\omega \ll \omega_0$, the occupation of a defect center can follow the AC signal frequency and the charge involved in the trapping or detrapping process will contribute to the measured capacitance. In contrast, when $\omega \gg \omega_0$, the same defect will not be able to follow the AC signal, and there will be no contribution from the defect to the capacitance.

On the contrary, the amplitude v_0 of the alternating signals also influences the density of the released charge carriers from the localized states in the depletion region. The process can be explained as follows. At zero applied voltage, the quasi-Fermi level E_{Fn} or E_{Fp} is constant in the bulk sample and the interface. With an applied AC signal, the quasi-Fermi level is modified by the varying potential. The levels close to E_{Fn} (or E_{Fp}) will cross the quasi-Fermi level, and the carriers occupying the traps in the vicinity of the intersection point can be released and contribute to the measured capacitance, say C_{ST} (ST stands for shallow traps). When the DC signal is large, the depletion region is extended to deeper states of the band gap and the released charge carriers from these states will contribute to the capacitance, as schematically shown in Figure 4.17. In other words, there will be a second capacitance C_{DT} (DT stands for deep traps) that is added to C_1, and the total capacitance of the device becomes $C_{ST} + C_{DT}$. The energetic and spatial distributions of the trap states in the band gap of the semiconductor will influence the charge released by the applied AC and DC voltages, and hence the capacitance of the device.

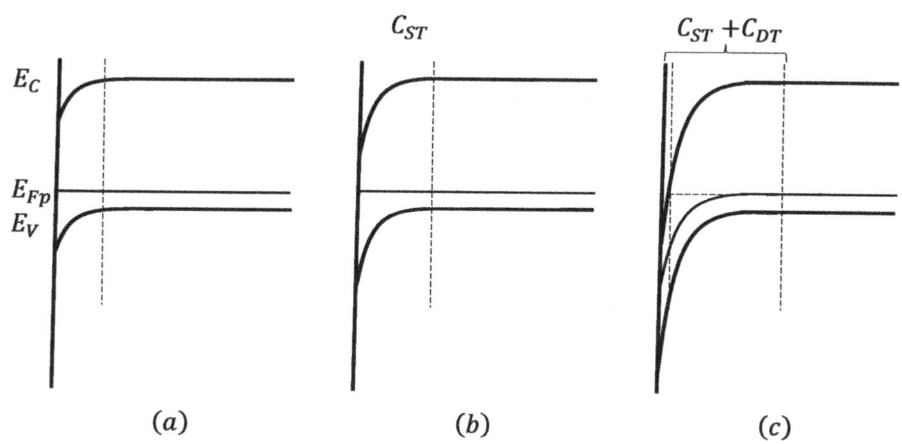

Figure 4.17. *Schematic representation of the capacitance related to the depletion region at the electrode/semiconductor contact with (a) no applied DC voltage, (b) small applied DC voltage and (c) large applied DC voltage*

Therefore, two factors can influence the depletion capacitance measurement: the frequency ω and the amplitude v_0 of the AC signals. Both factors are strongly linked to the depletion region formed at the interface of the semiconductor with the contact material (electrode or transport layer). Investigations of defects in organic semiconductors by capacitance measurements usually address these two aspects. The capacitance of the devices is usually studied as a function of applied DC voltage (or the *C–V profiling*) and as a function of frequency of applied AC voltage (or the *admittance analysis*). For the former study, the frequency of the AC signals is chosen to facilitate the thermal emission of charge carriers from trapping centers (low frequency) and kept constant during the device capacitance measurements as a function of applied DC voltage. For the latter study, the DC voltage is chosen to avoid strong current injection from the electrode, which may affect the thermal emission of carriers from defects. A reverse bias or a small forward bias is generally adopted for identifying the contribution of trapping charge to the measured capacitance. A complete study of the capacitance characterization usually combines both techniques and is carried out as a function of temperature to provide the carrier mobility and defect states in the semiconductor.

For a plate capacitor having two parallel electrodes of surface A and separated by a distance d, the *geometrical capacitance* is:

$$C_0 = \varepsilon_0 \frac{A}{d} \quad [4.70]$$

Therefore, as long as the relaxation time of the released charge carried from trapping centers is short as compared to the frequency of the applied AC signals, the measured capacitance is related to the depletion layer formed at the contact of the Schottky junction. It depends on the width x of the layer by:

$$C_{DR} = \varepsilon_0 \varepsilon_r \frac{A}{x} \quad [4.71]$$

In this expression, x is the depth of the depletion region with respect to the contact semiconductor/electrode ($x = 0$).

When the temperature of the sample is low or when the frequency of the applied AC voltage becomes high, the measured capacitance is related to the bulk of the semiconductor. It depends on the thickness d of the layer by:

$$C_B = \varepsilon_0 \varepsilon_r \frac{A}{d} \quad [4.72]$$

The transition between two processes is observed in the capacitance spectrum by a step of the capacitance when varying the temperature or the frequency of the AC voltage.

4.3.3.3. *Capacitance–DC voltage profiling*

In this section, the defects in devices are studied by the C–V measurements with the assumption that the frequency of the AC voltage is higher than that of the free-carrier relaxation. This technique provides information on defects located in or near the depletion region. The *depletion approximation* is assumed, that is, the depletion region is abrupt and contains no free charge carriers (Sze 1981). Furthermore, when the width of the depletion region, which is determined by the distance x from the contact interface, varies with the V_{DC} voltage, the charge carrier density is considered as constant in this region by this approximation. From equation [4.64], we have:

$$\left(\frac{1}{C_J}\right)^2 = \frac{2(V_{bi}-V_{DC}-2kT/|e|)}{|e|\varepsilon_0\varepsilon_r A^2 N} \qquad [4.73]$$

A plot of $(1/C_J)^2$ as a function of the applied voltage V_{DC} (called the *Mott–Schottky plot*) is a straight line, from which the carrier density N can be determined by the slope and the built-in voltage V_{bi} by the intercept with the x-axis. We obtain:

$$N = -\frac{2}{|e|\varepsilon_0\varepsilon_r A^2}\left[\frac{d\left(\frac{1}{C_J^2}\right)}{dV_{DC}}\right]^{-1} \qquad [4.74]$$

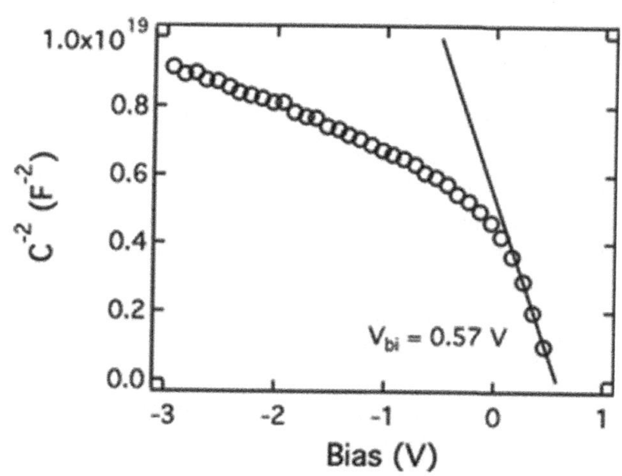

Figure 4.18. *The Mott–Schottky plot of a P3HT-based diode at temperature T =295 K, using a modulation frequency of 100 Hz. The extract built-in voltage is* $V_{bi} = 0.57\ V$. *(from Li et al. 2011, p. 1879)*

Figure 4.18 shows the Mott–Schottky plot obtained in an $ITO/PEDOT{:}PSS/P3HT/Al$ structure using a modulation frequency of 100 Hz at ambient temperature T = 295 K (Li et al. 2011). The built-in voltage V_{bi} is determined by the intercept of the straight line with the x (bias) axis.

For a non-uniform material, the carrier density is usually and more conveniently expressed by (Blood and Norton 1978):

$$N(x) = -\frac{C_J^3}{|e|\varepsilon_0\varepsilon_r A^2}\left(\frac{dC_J}{dV_{DC}}\right)^{-1} \qquad [4.75]$$

Using equation [4.71], the thickness of the depletion can be determined and then used with equation [4.75] to derive the profile of the carrier $N(x)$ concentration from point-to-point capacitance–DC voltage measurements. With the assumption of a constant trap density of traps, the depletion capacitance given by equation [4.71] should be corrected by taking into account the Debye screening length L_D, which represents the transition region at the depletion edge and is usually supposed to be negligible with regard to the thickness x of the layer (Johnson and Panousis 1971). With this consideration, the depletion capacitance is written as:

$$C_{DR} = \varepsilon_0\varepsilon_r\frac{A}{x+L_D} \qquad [4.76]$$

with:

$$L_D = \sqrt{\frac{\varepsilon_0\varepsilon_r kT}{|e|^2 N}} \qquad [4.77]$$

It should be noted that the exploitation of the results is based on expressions that are established within several assumptions that may not be valid for real devices (ideal depletion region, negligible contact resistance, no interface states). In particular, the potential obtained from the Mott–Schottky analysis of organic devices does not correspond to the built-in potential V_{bi} as in inorganic semiconductors (Mingebach et al. 2011). Moreover, the measured carrier density N, which represents the density of defects in the depletion region, is indeed the density at the depletion region edge and not in the bulk of the semiconductor. The depletion approximation holds true for trap centers of low-level energy or shallow traps, by assuming that the occupancy of deep traps is not significantly perturbed by the variation of the applied DC voltage, and consequently does not affect the width of the depletion region. In semiconductors of high deep trap concentration, the trap measurements with the depletion approximation may then lead to overestimate the defect concentration, as explained previously.

A technique called *drive-level capacitance profiling* (DLCP) (Michelson et al. 1985) can be used to take into account the influence of the deep traps on the capacitance measurements. This technique, applied for diodes in reverse bias or with small forward bias, is based on the determination of the additional terms due to the effect of the amplitude of the AC signals, and consequently of the deep traps, on the capacitance whose expression can be written as:

$$C = \frac{dQ}{dV} = C_0 + C_1 dV + C_2\,(dV)^2 + \cdots \qquad [4.78]$$

In this expression, C_0 is the junction capacitance for small amplitude AC voltage, which is given by (Heath et al. 2004):

$$C_0 = \frac{\varepsilon_0 \varepsilon_r A \rho_e}{\varepsilon_0 \varepsilon_r F_e + x_e |\rho_e|} \qquad [4.79]$$

where x_e, F_e and ρ_e are the distance from the barrier interface, the electric field and the charge density at the point x_e, where the quasi-Fermi level lies at the energy level E_e such that the carriers from the states located near it can be emitted, that is, $E_e = E_{Fp} - E_V$ in a P-type semiconductor. The *characteristic angular frequency* ω_0 of the alternating voltage, which corresponds to the limit of the release of trapped charge carriers, is related to the thermal emission rate (e_n or e_p). For a P-type semiconductor, the emission rate of these carriers is given by:

$$\frac{1}{\tau_p} = \omega_0 = e_p = N_V v_{th} \sigma_p exp\left(-\frac{E_e}{kT}\right) \qquad [4.80]$$

where N_V is the effective valence density of states, v_{th} is the thermal velocity, and σ_p is the hole capture cross-section. Assuming negligible temperature dependence, the emission rate can be written as:

$$e_p = \gamma \sigma_p T^2 exp\left(-\frac{E_e}{kT}\right) \qquad [4.81]$$

where $= N_V v_{th} T^{-2}$.

Depending on the location of the trap states with respect to the quasi-Fermi level within the depletion region, a state can answer to the applied AC signals or not. Therefore, the demarcation energy E_e delimits the states that can and those that cannot follow the AC signals. Traps below this level can change their energetic state and contribute to the capacitance charge while those located above it cannot and will not contribute to the capacitance. From equation [4.81], we can write:

$$E_e = E_{Fp} - E_V = -kT \ln\left(\frac{\omega_0}{\gamma \sigma_p T^2}\right) \qquad [4.82]$$

The coefficient C_1 can be derived from equation [4.78] by writing the change in depletion charge dQ in response to dV as $dQ = A\rho_e dx$, that is, the charge comprised between x_e and $x_e + dx$. We obtain:

$$C_1 = -\frac{1}{2}\frac{\varepsilon_0\varepsilon_r A\rho_e^2}{(\varepsilon_0\varepsilon_r F_e + x_e|\rho_e|)^3} \quad [4.83]$$

It is then possible to plot the variation of the charge carrier density N_{DL} (DL for drive level) in the depletion region, which includes released charges from deep traps, as a function of the spatial parameter $W = \frac{\varepsilon_0\varepsilon_r A}{C_0}$. From equations [4.79] and [4.83], the density N_{DL} is determined as (Michelson et al. 1985):

$$N_{DL} = \frac{\rho_e}{|e|} = -\frac{C_0^3}{C_1} \times \frac{1}{2|e|\varepsilon_0\varepsilon_r A^2} \quad [4.84]$$

The density is directly linked to the density of states in the material since the volume charge density can be written as (Heath et al. 2004):

$$\rho_e = |e| \int_{E_{F0}}^{E_V + E_e} g(E, x_e)dE \quad [4.85]$$

where E_{F0} is the Fermi level in the bulk of the material.

Therefore, the measurement of the density N_{DL} makes it possible to determine the energetic (density of trap states as a function of trap energy in the band gap) and spatial (density of trap states as a function of position or distance from the physical contact with the electrode) distribution of defects in the semiconductor. Practically, the DLCP profile is obtained by plotting N_{DL} as a function of the profile distance $\langle x \rangle = \frac{\varepsilon_0\varepsilon_r}{C_0}$. The profile distance can be written from equation [4.79] as:

$$\langle x \rangle = x_e + \frac{\varepsilon_0\varepsilon_r F_e}{|\rho_e|} \quad [4.86]$$

It depends practically on the position x_e of the defect emission since the second term of equation [4.86] can be considered as a constant when the applied DC voltage changes.

As an example, Figure 4.19 shows the plot of the capacitance of an $ITO/NPB/Al$ structure as a function of AC signal amplitude (Pang et al. 2019). NPB is used here as an active organic material of the diode. The DC applied voltage is set at 1.5 V, and the frequency is set at 200 Hz. The capacitances C_0 and C_1 are obtained by fitting the experimental curve and are used to determine the carrier density N_{DL} by equation [4.84].

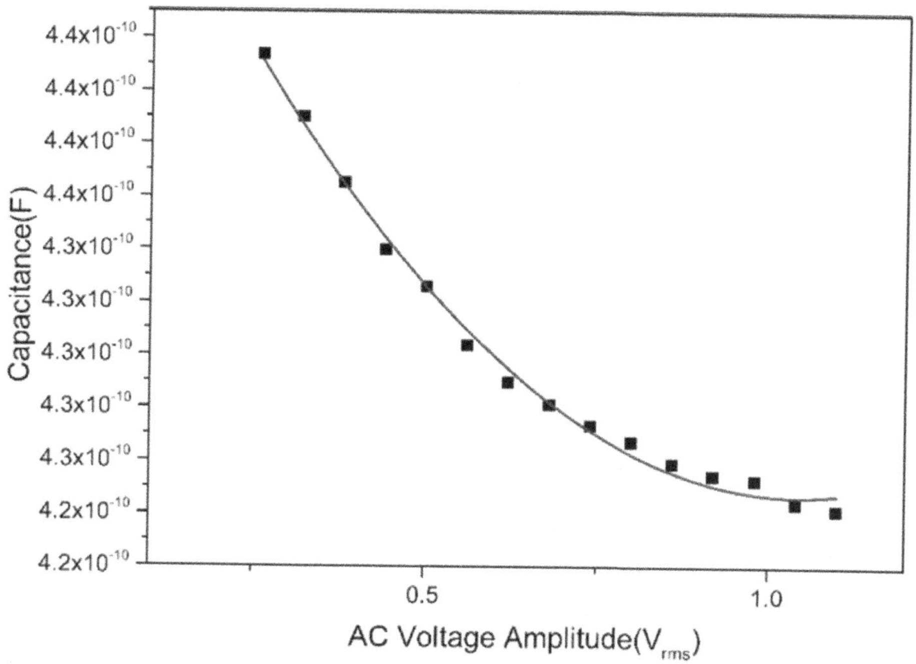

Figure 4.19. *Drive-level capacitance profiling showing the variation of junction capacitance with AC bias amplitude. The red line shows a quadratic fit, which yields C0 and C1 (from Pang et al. 2019, p.165)*

The DLCP technique is currently used for characterizing defects of inorganic semiconductors but is much less popular in organic field. The main difficulty of the capacitance determination of organic devices relies on the lack of accuracy of the measurements, which is usually very small (see, for instance, Figure 4.19). As shown previously, using the Mott–Schottky analysis, it is possible to extract the carrier density and the built-in voltage from the C–V profiling of organic diodes. Furthermore, by applying equation [4.75] and by measuring point-by-point the capacitance, it is possible to obtain the spatial representation of defects in these devices (Carr and Chaudhary 2013). Further complementary investigations by capacitance measurements as a function of frequency make it possible to verify and confirm the validity of the depletion profiling technique.

4.3.3.4. *Admittance spectroscopy*

Admittance measurements consist of recording the capacitance of the device as a function of frequency of the applied AC signals, which are superimposed on the applied DC voltage. The use of a DC applied voltage is normally not necessary to investigate defects by admittance analysis. However, the depletion region created by the DC voltage may be used by its variation to distinguish between defects from the interface or from the bulk in devices.

Before examining the effects of defects on the device capacitance $C(f)$ characteristic, we shall recall some of the basic representations currently used in the admittance analysis.

4.3.3.4.1. Basic circuits

The admittance of a device given in equation [4.69] can be represented by a parallel circuit comprising a resistance R and a capacitance C, which describes the device response under an applied AC voltage (Figure 4.20a). The impedance of such a circuit is:

$$Z(\omega) = \frac{R_P}{1+j\omega R_P C_P} \quad [4.87]$$

At low frequencies, the impedance of the capacitance becomes infinite and $|Z| \rightarrow R_P$, which corresponds to a plateau on the frequency scale. At high frequencies, the impedance of the capacitance vanishes and $|Z| \rightarrow 0$. The capacitance function $C(\omega)$ is constant and equals C_P.

In real devices, due to the connection of the electrodes to the measurement instruments, a series resistance R_S should be added to the RC circuit (Figure 4.18b). In organic devices, the series resistance is generally small (less than 100 Ω) as compared with the bulk resistance. The impedance of such circuit is:

$$Z(\omega) = R_S + \frac{R_P}{1+j\omega R_P C_P} \quad [4.88]$$

At low frequencies, the impedance becomes resistive and $|Z| \rightarrow R_S + R_P$. At high frequencies, the impedance of the parallel circuit vanishes and $|Z| \rightarrow R_S$. Because of the series resistance, a potential drop of the applied voltage over R_P takes place and the capacitance function $C(\omega)$ shows a plateau at low frequencies, which is lower than the parallel capacitance C_P.

The expression of the capacitance of the circuit is changed to:

$$C(\omega) = \frac{C_P R_P^2}{(R_S+R_P)^2+(\omega R_P C_P)^2} \quad [4.89]$$

which reaches the following limit when $\omega \rightarrow 0$:

$$C(\omega \rightarrow 0) = \frac{C_P R_P^2}{(R_S+R_P)^2} \quad [4.90]$$

The capacitance plateau coincides with the value of the parallel capacitance when the series resistance R_S is small compared with the parallel resistance R_P.

The capacitance as a function of frequency for the two circuits $(R_P C_P)$ and $(R_S, R_P C_P)$ is shown in Figure 4.20c and d.

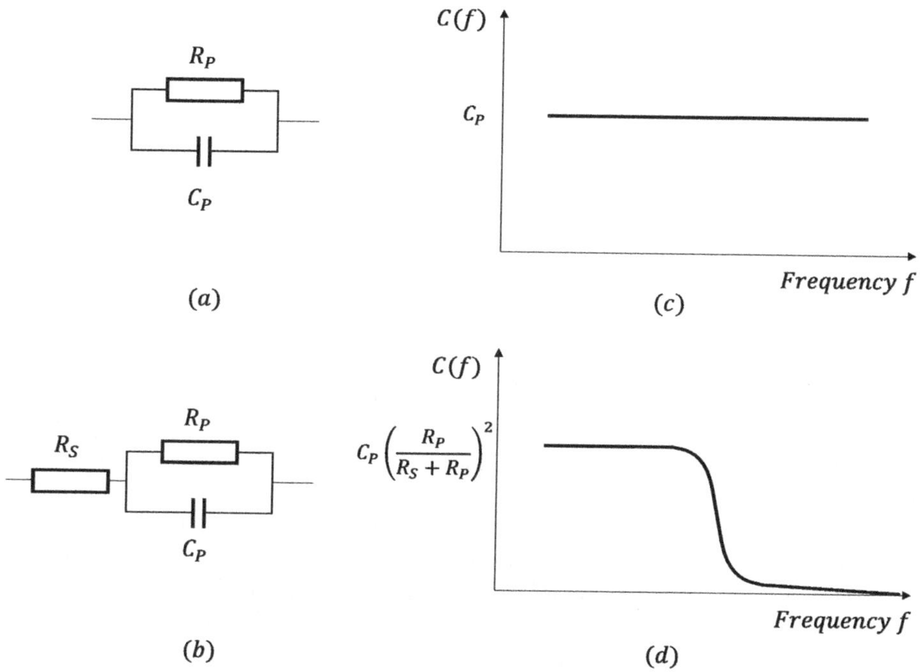

Figure 4.20. *Basic equivalent circuits and their capacitance function as a function of frequency: (a) RC parallel circuit, (b) $R-RC$ series–parallel circuit, (c) $C(f)$ spectrum of the RC parallel circuit and (d) $C(f)$ spectrum of the $R-RC$ parallel circuit*

Another method of representation of the impedance data consists of plotting the imaginary part $X(\omega)$ as a function of real part $R(\omega)$ of the impedance (the Nyquist plot). In this representation, the angular frequency is not clearly identified and only its limit values ($\omega \to 0$ and $\omega \to \infty$) can be located by the corresponding values of the real part $R(\omega)$. For a simple $(R_P C_P)$ circuit, the Nyquist plot is a semi-circle centered on the real axis and of diameter R_p. When $\omega \to \infty$, the capacitance is small and shunts the resistor: the semi-circle passes through the origin. For the $(R_S, R_P C_P)$ circuit, the semi-circle is shifted towards the high values of the $R(\omega)$ axis. The shifted distance from the origin corresponds to the series resistance R_S (Figure 4.21b).

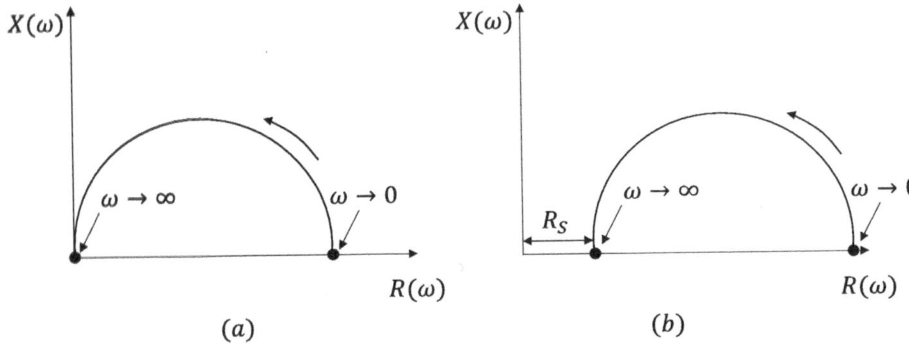

Figure 4.21. *The Nyquist plots of the basic equivalent circuits: (a) RC parallel circuit and (b) R – RC series–parallel circuit. Arrows indicate the direction of increasing angular frequency*

4.3.3.4.2. Complex circuits

An electronic device such as organic light-emitting diode or organic solar cell is composed of several layers, each of which play a specific role in the physical processes that govern the operation of the device. It can be understood that there may be several equivalent circuits corresponding to different processes, including charge carrier trapping, which compose the overall equivalent circuit of the device. For instance, in typical organic devices, the transport layers (ETL for electrons and HTL for holes) are incorporated to facilitate the carrier injection from the electrodes into the active layer. Each of these layers has a generally low conductivity and a dielectric constant and can be represented by a parallel RC circuit. These two cells are added to the equivalent circuit for studying the device response under an AC applied voltage.

In some configurations, the equivalent circuit of a device can be simplified to a basic form (RC circuit or R-RC), but in most cases, a complex equivalent circuit should be built from the analysis of the possible physical processes of the device operation together with the possible location of the layer in which they occur. Each of them is represented by an equivalent circuit and then, these circuits are brought together to obtain the equivalent circuit of the device. In particular, the effects of the defects will be described using additional electronic components which will be incorporated to the overall circuit.

In the equivalent circuits, the electronic components (capacitance and resistance) are supposed to be ideal, and their combination judiciously chosen makes it possible to correctly fit the experimental spectra and allows them to identify and to understand the physical processes occurring in the devices. However, it happens that the use of these components may lead to a dead end when computing the response of the circuit. Such a problem is believed to be due to the morphology of the surface of layer(s), which creates imperfect interfaces between them. The density of charge of the capacitance formed at such an interface would not be uniform and the interface will behave as a leaking capacitor. To describe this behavior, a new component called a *constant phase element* (CPE) can be used to compose the equivalent circuit (Macdonald 1987). Its impedance function is given by:

$$Z_{CPE}(\omega) = A(i\omega)^{-\alpha} \qquad [4.91]$$

Depending on the value of the exponent α, a CPE behaves as an ideal capacitor ($\alpha = 1$) or as an ideal resistor ($\alpha = 0$). The intermediate case with $\alpha = 0.5$ corresponds to a *Warburg-like impedance*. In practice, for fitting the admittance spectra of devices using an equivalent circuit, the exponent α is considered as a variable, which allows for the adjustment of experimental curves. Although the use of CPE is currently adopted for simulation of the admittance spectra in organic devices, the physical meaning and its interpretation remain unclear and this needs to be checked by other considerations and experiments.

In order to consider the contribution of a single defect in organic devices through admittance measurements, a series RC cell, for a single trap state, is added to the equivalent circuit of the device free of defects (Losee 1975). As previously mentioned, the latter circuit should include, in principle, all the representative elements of the device in operation, that is, the physical elements (contact, transport, bulk layers) and the process elements (injection, recombination, tunneling of charge carriers) which can be identified. However, a complete equivalent circuit cannot be used for simulation of the AC response of the device because its mathematical treatment becomes too complex. Therefore, a simplified circuit, which can reflect the dynamic response of the device and its structure properly, should be used for studying the defects in organic devices.

Such a simplified equivalent circuit may be composed of an $R_P C_P$ parallel circuit connected to a series resistance R_S as shown in Figure 4.22b (Burtone et al. 2012). In this representation, the parallel circuit and the series resistance include the different contributions from the physical elements and processes that can be identified in the device free of defects. Theoretically, each element and each process can be separately examined by modifying the device structure or the experimental conditions for the impedance measurements. However, for a defect study, this simple equivalent circuit can be used together with an additional circuit representing the contribution of trap centers to the charge of the device under an AC applied voltage, which will be examined in the next section.

Figure 4.22 shows two examples of the Nyquist plot of a simple and a complex device. For organic devices, the semi-circles in Figure 4.22b can correspond to the impedance of two regions in the semiconductor layer. The small one on the left is identified in the high-frequency range and can be attributed to the bulk process of the semiconductor. It should noticeably increase with increasing temperature at a constant applied voltage and would not change with the increasing applied voltage at a constant temperature. The large semi-circle on the right is identified in the low-frequency range and can be attributed to the depletion layer formed at the electrode–semiconductor contact. It should noticeably increase with the increasing applied voltage at a constant temperature and normally decreases with an increasing temperature at a constant voltage. For each studied device, it would be necessary to investigate the effects of the temperature and the applied voltage on the impedance spectra to extract the information on the defect parameters.

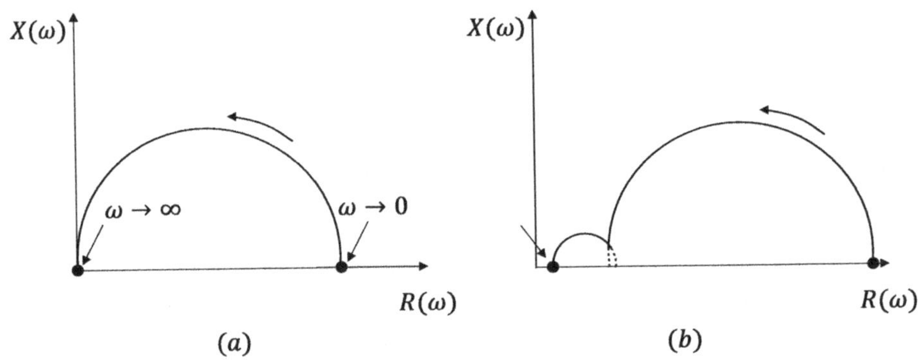

Figure 4.22. *Nyquist plot for: (a) a perfect device with negligible contact resistance, (b) a complex device or device with complex defect processes*

4.3.3.4.3. Effects of defects in the admittance of organic devices

The measured capacitance of a device includes contributions from the semiconductor free carriers and from defects that stem from two different processes: charge emissions in the depletion region by the DC voltage and charge emissions in the allowed frequency range of the AC signals. The capacitances associated with the processes are denoted as C_D (dielectric capacitance), C_J (junction capacitance) and $C_t(\omega)$ (trap frequency-dependent capacitance).

By choosing the appropriate experimental conditions (high temperature and low frequency), it is possible to measure only the contribution to the capacitance by the defects, that is, excluding the capacitance C_D, which is due to the free charge carriers in the semiconductor (freeze out of free carriers) (Lee et al. 2005). This capacitance is independent of applied voltage, and its value is close to the geometrical capacitance. In this case, the device capacitance can be considered as a sum of two capacitances: one frequency-dependent capacitance $C_t(\omega)$ and one frequency-independent capacitance C_J.

For a single trap level, the capacitance can be expressed as (Jasenek et al. 2000):

$$C(\omega) = C_J + \frac{C_{lf} - C_j}{1+\omega^2\tau^{*2}} \quad [4.92]$$

where C_{lf} is the low-frequency capacitance plateau due to the released charge carriers from trapping centers, and τ^* is the time constant. If the defect density is small compared with the acceptor density N_A for a P-type semiconductor, τ^* is close to the value of ω_0 which is related to the characteristic angular frequency by $\omega_0\tau_0 = 1$.

From equation [4.92], the capacitance spectrum presents a plateau at a high frequency, which corresponds to the depletion capacitance C_J. At a low frequency, a capacitance plateau of value C_{lf} is presented, which represents the contribution of the charge carriers released from traps.

The typical admittance capacitance of an organic device as a function of frequency of the applied AC voltage is shown in Figure 4.23, and the effect of the defects can be observed by an increase of the capacitance in the low-frequency region. The charge carriers are released from the trapping centers and contribute to the capacitance whenever the frequency ω of the AC signals is lower than the carrier relaxation time τ_n or τ_p. The characteristic angular frequency ω_0 corresponds to the limit of the release of trapped charge carriers related to the thermal emission rate (e_n or e_p). With the assumption of a low trap density in the junction, the characteristic frequency ω_0 is the inflexion frequency of the electronic transition. It can be determined by plotting the differentiation function of the capacitance $-\omega\frac{dC}{d\omega}$ as a function of frequency, which shows a maximum at a frequency ω_0. This plot is then used to determine the thermal emission rate (e_n or e_p) of the charge carriers, then the defect parameters (the activation energy and the capture cross-section) by studying the temperature dependence of the frequency ω_0 using an Arrhenius plot.

At a very high angular frequency, the carrier emission rate is too low, and the charge contribution from the defect states becomes negligible. The capacitance of the device presents a plateau which corresponds to its geometrical value C_0.

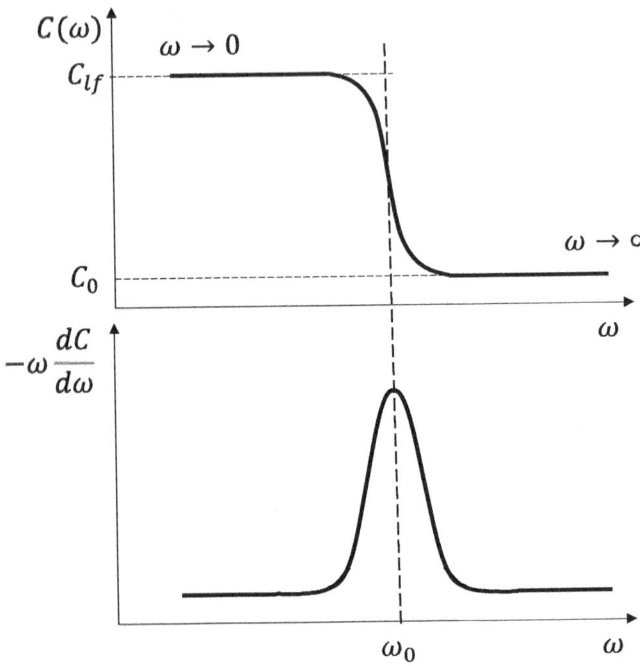

Figure 4.23. *Schematic representation of the capacitance and its differentiation as a function of angular frequency in thermal equilibrium condition (T = constant)*

As an example, the admittance measurements of an $ITO/MDMO\text{-}PPV:PCBM/Al$ cell at different temperatures are given in terms of a capacitance and a differential capacitance as a function of frequency in Figure 4.24 (Dyakonov et al. 2001). It can be seen that the $C(\omega)$ spectrum shows a step, which shifts towards a higher frequency when the temperature increases. This step corresponds to a peak in the differential capacitance spectrum, which also shifts towards higher frequencies at higher temperatures. The results indicate the presence of trap states in the organic blend with an energy activation in a range of 24–34 meV, which corresponds to a shallow trap level.

A method for studying defect distributions of a device by capacitance measurements is examined by Walter et al. (1996). It consists of evaluating the contribution of an energy distribution of defects $N_t(E)$ at a given position in the depletion region to the capacitance, then an integration of the results in space and energy for defects at the Fermi level is performed with the consideration of the limit frequency beyond it, all defects cannot follow and will not contribute to the capacitance.

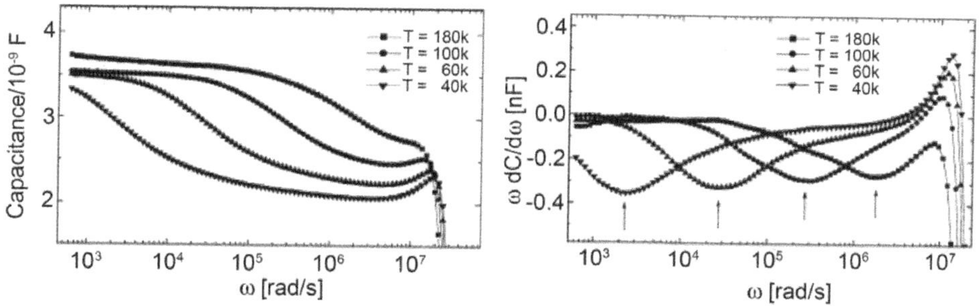

Figure 4.24. *Capacitance (left) and differential capacitance (right) of an ITO/MDMO-PPV:PCBM/Al cell measured at different temperatures (from Dyakonov et al. 2001, p. 103)*

The defect distribution at a demarcation energy $E_e(\omega)$ or E_ω can be determined by calculating the derivative of the measured capacitance $C(\omega)$ by:

$$N_t(E_\omega) = -\frac{V_{bi}}{|e|W}\frac{dC(\omega)}{d\omega}\frac{\omega}{kT} \quad [4.93]$$

where V_{bi} is the built-in voltage in the junction, and W is the width of the intrinsic depletion layer (V_{bi} and W are usually determined by the capacitance–voltage measurements).

From equation [4.82], the demarcation energy is rewritten as:

$$E_\omega = kT \ln\left(\frac{\gamma\sigma_p T^2}{\omega}\right) \quad [4.94]$$

It is proportional to $\ln(\omega)$ and the conversion of the frequency to energy can be made using this expression. From equations [4.93] and [4.94], the plot of $N_t(E_\omega)$ is made possible from the experimental capacitance $C(\omega)$ curve and describes the defect distribution in the band gap of the semiconductor.

As an example, the Walter model has been used to study the defect energy distribution in the $CH_3NH_3PbI_3$ perovskite solar cells (Duan et al. 2015), shown in Figure 4.25. The distribution has two components centered at 0.167 and ~0.3 eV, respectively. The main component can be fitted with a Gaussian function and the corresponding defect density is $\sim 10^{16} cm^{-3}$. These defects are assigned to iodine interstitials.

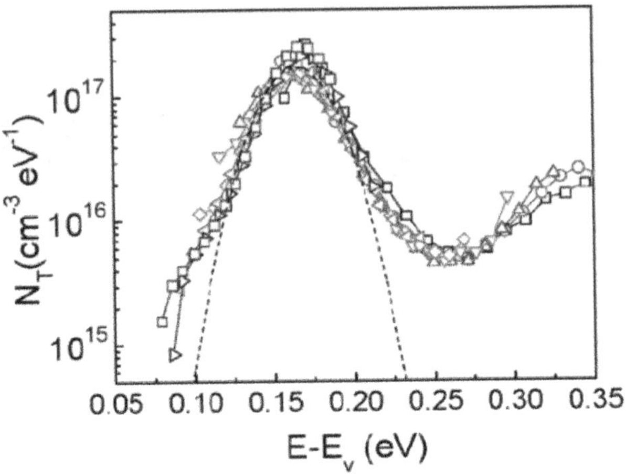

Figure 4.25. *The defect energy distribution of $CH_3NH_3PbI_3$ perovskite (from Duan et al. 2015, p. 112). For a color version of this figure, see www.iste.co.uk/nguyen/defects.zip*

4.3.3.4.4. Equivalent circuits of devices containing high defect density

The equivalent circuit of a device is used to describe its response to the applied AC signals in the admittance characterization. A simplified circuit composed of an R_PC_P parallel circuit and a series resistance R_S is commonly used to fit the measured data with satisfactory results in devices with a negligible trap density. In presence of a high trap density, the circuit is modified by adding a complementary circuit representing the contribution of the traps to the overall response of the device. For single trap levels in a Schottky-barrier diode, the equivalent circuit proposed by Losee (1975) is composed of two sub-circuits as shown in Figure 4.26a. The R_iC_i cells represent discrete traps levels and are such that $R_iC_i = \tau_i$ is the time constant of the corresponding cell. The capacitance C_∞ represents the contribution of the free charges (in the conduction or valence band).

As the trap states are distributed in the band gap of the semiconductor, the equivalent circuit should have an infinite number of elementary parallel R_iC_i cells, and the representation becomes mathematically too complex. It is much simpler to replace the elementary cells by an impedance Z_T representing the distribution of traps (Bisquert 2008). The equivalent circuit of devices with traps in Figure 4.26b can be then applied to interpret the experimental capacitance spectra.

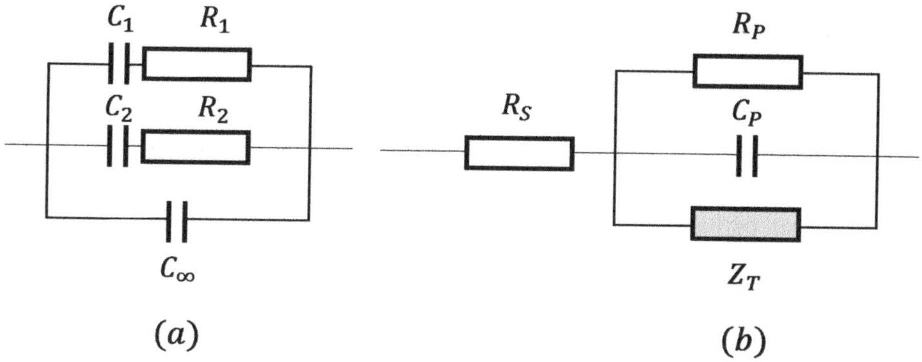

Figure 4.26. *Equivalent circuits of devices containing high trap densities: (a) Losee model for a Schottky-barrier diode and (b) simplified circuit for devices with high density of traps*

For fitting the experimental capacitance spectra of devices, the commercial program Zview can be used. The elements of the simplified equivalent circuit in Figure 4.18, that is, the series resistance R_S, the parallel resistance R_P, and the parallel capacitance C_P, will be obtained. The effects of the trap states are studied by comparing the experimental spectrum with the simulated capacitance spectra using the circuit in Figure 4.26b. Simulations of both exponential and Gaussian trap distributions have been performed to investigate the defects in organic devices.

The trap capacitance $C(\omega)$ is calculated by integration of their energy distribution $g_t(\omega)$ with occupancy $f(E)$ and assuming the Boltzmann approximation for the trap occupation. The trap distribution function $g_t(\omega)$ can be Gaussian or exponential and are given by equations [2.5] and [2.6]. For a P-type semiconductor, the density of occupied hole traps is given by:

$$p_t(E) = \int g_t(E)[1 - f(E)]dE \qquad [4.95]$$

The kinetics equations for trapping and detrapping processes are (Duijnstee et al. 2020):

$$\frac{\delta p}{\delta t} = -\frac{\delta p_t}{\delta t} = \int g_t(E)\frac{\delta f(E)}{\delta t}dE \qquad [4.96]$$

$$\frac{\delta f(E)}{\delta t} = \beta p[1 - f(E)] - \varepsilon f(E) \qquad [4.97]$$

where β and ε are the time constants for hole capture and release respectively.

Under a small applied AC signal dV, the charge density of the device varies by a quantity dQ. The capacitance is expressed as a ratio of the charge variation dQ to the voltage variation dV:

$$C_T(\omega) = \frac{dQ}{dV} = \frac{|e|}{dV}(dp + dp_t) = \frac{|e|}{dV}\left(dp + \int g_t(E)[1 - df(E)]dE\right) \quad [4.98]$$

For a small perturbation:

$$df(E) = \frac{1}{p}\frac{f(E)[1-f(E)]}{1+i\omega/\omega_t}dp \quad [4.99]$$

where ω_t is the trap frequency and is given by:

$$\omega_t = \frac{\varepsilon}{1-f(E)} \quad [4.100]$$

Using equations [4.98] and [4.99], the trap capacitance can be written as:

$$C_T(\omega) = \frac{|e|^2}{kT}\int \frac{f(E)[1-f(E)]}{1+i\omega/\omega_t} g_t(E)dE \quad [4.101]$$

With a known trap distribution function, the trap capacitance can be calculated and be compared with the measured data to identify the contribution of the charge emission processes (Pahner et al. 2013).

4.3.3.5. *Impedance measurements of organic semiconductor and devices*

Compared with other electrical characterization techniques, the impedance measurements of devices do not need a sophisticated set-up and are relatively simple to carry out. The method is largely used to study electrical transport processes in organic devices. In an ideal device, it is possible to determine the trap parameters such as the activation energy, the capture cross-section and the density of states. It should be emphasized, however, that the interpretation of the impedance spectra may lead to confusion and consequently, should be undertaken with care. Indeed, several physical processes may simultaneously occur during the device impedance measurements, under applied DC and AC voltages. According to Kneisel et al. (2000), the following mechanisms may be encountered in an admittance spectrum: (i) dielectric relaxation of the semiconductor (geometrical capacitance by freeze-out of the free carriers), (ii) formation and expansion of the depletion region (frequency-independent capacitance), (iii) trapping and detrapping of charge carriers (frequency- and temperature-dependent capacitance), (iv) trap distribution effect and (v) decrease of capacitance at the cut-off frequency due to a series resistance. Several of these mechanisms may occur during the experiment leading to similar admittance signatures and make ambiguous the interpretation. It is, of course, possible to separately investigate a given mechanism by choosing appropriate experimental conditions (for instance, by setting temperature and the frequency range) when performing the measurements. The analysis of the data is time consuming, and in some cases, complementary information on the semiconductors is required, meaning that further experiments should be performed in order to exploit the admittance results. In addition to the mentioned mechanisms, the analysis of admittance spectra for equivalent circuit modeling should take into account the back contact of the Schottky diode, which affects the device characteristics, and may lead to a misinterpretation of the trap parameters (Eisenbarth et al. 2010). In the presence of a back contact potential carrier, the conductance and the capacitance of the device change and the measured characteristic frequency, especially in the high-frequency range, may not correspond to genuine traps in the device. To characterize the back contact, a parallel $R_C C_c$ cell is connected in series to the simplified equivalent circuit shown in Figure 4.26b. An example of a complex equivalent circuit model, and its use to interpret the physical processes from analysis of the measured data can be found in the article of Proskuryakov et al. (2007). It is evident that the previous remarks made on using IS for characterizing defects in inorganic devices can be applicable to a wide range of organic ones. It should be kept in mind, however, that as the carrier mobility in organic materials is much lower than in classical semiconductors, the measurements should be carried out with adapted experimental conditions, namely, the frequency range used, in order to obtain reliable results (Mingebach et al. 2011).

In the field of organic device characterization, IS has been largely applied to investigate defects in conjugated polymers, small molecules and hybrid organic–inorganic materials. As previously mentioned, only some typical examples of defect measurements will be given below.

For conjugated polymers, PPV has been investigated in diodes of the structure $metal/PPV/metal$ (Nguyen and Tran 1995) in the frequency range from 1 mHz to 500 kHz. The conductance and capacitance of the devices recorded as a function of temperature are modeled by using an equivalent circuit from the Maxwell–Wagner –Sillars representation, which suggests two defect levels centered at 0.4 and 0.7 eV from the band edge. The observed relaxation mechanisms are assigned to trapping by centers localized in the bulk and the polymer–metal interface respectively.

In MEH-PPV-based diodes of structure $Au/MEH\text{-}PPV/Ca$, the impedance of the devices measured from 100 Hz to 1 MHz is modeled by a parallel RC cell (Campbell et al. 1995). The measured trap density at the metallic contact is estimated to be of the order of $10^{16} cm^{-3}$. Further impedance study of MEH-PPV diodes with different electron injecting cathodes (Ca, Al, Au) in the frequency range from 20 Hz to 1 MHz shows the influence of trapped charges at the interface and injected charges from the electrode to the polymer (Shrotriya and Yang 2005). The equivalent circuit used in this analysis is an $R_P C_P$ parallel circuit and a series resistance R_S, which suggests that different combinations of capacitance and resistance can be used for simulating the impedance data obtained from the same polymer-based devices.

Defects in conjugated polymer P3HT are investigated by a modified capacitance–voltage measurement technique based on a forward bias of the device of the structure $ITO/P3HT/Al$ (Ray et al. 2014). In the classical technique, the C–V measurements are performed in the reverse bias and the Mott–Schottky analysis makes it possible to determine the (shallow) trap density of the semiconductor. With an applied voltage in the forward bias and a favorable energetic configuration of the device, both deep and shallow traps can be revealed in the C–V spectrum with two apparent capacitance peaks. The Mott–Schottky plot shows two slopes corresponding to the reverse and forward biases of the device, from these built-in potentials are extracted and used for determining the trap levels. The density of traps is found to be $2 \times 10^{15} cm^{-3}$ and $6 \times 10^{15} cm^{-3}$ for shallow and deep trap levels respectively. Their energetic levels are determined from the capacitance peaks and are localized at 0.35 eV for shallow traps and 1.3 eV for deep traps. Defects in P3HT:PCBM blends of different weight ratio are studied by both C–V and C–f measurements using devices of structure $ITO/PEDOT{:}PSS/P3HT{:}PCBM/Ca/Ag$ from 10 Hz to 500 kHz (Boix et al. 2009). The defect density derived from the Mott–Schottky analysis is $\sim 1 - 2 \times 10^{16} cm^{-3}$, and fitting results using a Gaussian distribution indicate a maximum energy of 0.38 eV. The defect parameters appear independent or weakly dependent on the blend composition suggesting that they belong to the polymer. It should be noted that the IS is frequently used to characterize the defects in P3HT and P3HT:PCBM blends, and there is a large discrepancy of the trap parameters results. This can be partly explained by the measurement conditions (Li et al. 2011) and also by the quality of the P3HT polymers, which is strongly dependent on the synthesis conditions, even when they are provided by commercial companies.

For small-molecule materials, the impedance spectrum performed on diodes of structure *ITO/pentacene/Al* from 100 Hz to 20 MHz as a function of DC applied voltage shows two distinct semicircles in the Nyquist plot, which are related to the charge carrier release from the bulk (high-frequency component) and the depleted region (low-frequency component) respectively. The density of traps is estimated to be of the order of $10^{17} cm^{-3}$. Further investigations of pentacene and CuPc-based diodes with structure *Al/pentacene/Au* and $Al/CuPc/Au$ are carried out by capacitance–voltage and capacitance–frequency measurements (Sharma et al. 2013). A Gaussian distribution of trap states can be determined with a trap level at 0.54 and 0.52 eV for pentacene and CuPc respectively. The trap density in both materials is in the range of $10^{16} - 10^{17} cm^{-3}$.

In Alq3 devices of structure $ITO/Alq3/Al$, impedance measurements are carried out from 20 Hz to 1 MHz and between 150 K and 320 K (Jeong et al. 2002). Analysis of the spectra makes it possible to determine the activation energy of $\sim 0.2\ eV$ of the trapping centers. Doped Alq3 with F4-TCNQ 4% shows a density of states above the Fermi level of the order of $10^{16} cm^{-3}$ and a strong dependence on the frequency of the capacitance in reverse bias (Ray and Narasimhan 2008). The involved traps are assigned to deep hole traps at energies above E_F.

Defects in hybrid perovskites $CH_3NH_3PbI_3$ are studied by measuring the capacitance–frequency characteristic of devices of structure $ITO/PEDOT{:}PSS/CH_3NH_3PbI_3/PCBM/C_{60}/BCP/Al$ at ambient temperature (Shao et al. 2014). The density of trap states is determined by applying Walter's model giving a density of the order of $10^{17} - 10^{19} cm^{-3}$ and an energy of $\sim$ 0.40 eV. We note that the value of the trap density is high for perovskite materials used in this study. Passivation of the perovskite by deposition of a PCBM layer over its surface followed by thermal annealing at $100°C$ makes it possible to reduce the trap density to a reasonable value in the range of $10^{16} - 10^{17} cm^{-3}$. In the work of Duan et al. (2015), the $C(f)$ characteristics are recorded at various temperatures and in devices of different architectures $FTO/spiro\text{-}OmeTAD/Au$, $FTO/TiO_2/spiro\text{-}OmeTAD/Au$ and $FTO/TiO_2/CH_3NH_3PbI_3/spiro\text{-}OmeTAD/Au$ in order to identify the true capacitance junction by comparison of the $C(f)$ spectra. The junction capacitance is then resolved and by using the same analysis technique as in the previous group, the trap parameters are determined to be $\sim 10^{16} cm^{-3}$ for the density and two energy Gaussian distributions at levels of $0.167\ eV$ and $\sim 0.3\ eV$. The shallow traps are assigned to iodine interstitials, which are non-radiative recombination defects in perovskite.

In perovskite-based devices, incorporation of extrinsic ions to the crystal may improve substantially their performance by increasing the carrier diffusion length. For instance, the diffusion length in $CH_3NH_3PbI_3$ is $\sim$ 100 nm and becomes $\sim$ 1 μm in $CH_3NH_3PbI_{3-x}Cl_x$ upon incorporation of chlorine in the perovskite (Chen et al. 2015). Using IS, the defects in perovskite devices with and without chlorine are determined and show that shallower trap levels ($0.21\ eV$) introduced by chlorine in the whole device is at the origin of the improved carrier extraction, and hence of the enhancement of the device performance. Here, the same remark on the quality of the material, as in the case of P3HT conjugated polymer, can be made on the discrepancy of the trap measurement results in perovskite-based devices. The synthesis conditions and the structure engineering of perovskite have a strong influence on the defect formation, in particular dislocations and grain boundaries, which needs treatments for obtaining reliable quality materials.

Although IS is acknowledged as an efficient and practical method to detect and characterize defects in semiconductor devices, it has several deficiencies such as the failure to distinguish between electron and hole trapping centers, or the difficulties for precisely determining the defect parameters even with complex mathematical treatments. A more sensitive characterization of defects exploiting the capacitance variation of devices is the DLTS, whose principle is explained in the following section.

4.3.4. *Deep-level transient spectroscopy (DLTS)*

DLTS introduced by Lang in 1974 is a popular technique for characterizing defects in semiconductors and devices (Lang et al. 1974). According to the author, it is "sensitive, rapid and straightforward to analyze". The advantages of the technique also cover the possibility to distinguish between the minority and majority-carrier traps, and to provide information about shallow and deep traps, which are not always attainable by other analysis techniques.

4.3.4.1. *Basic principle of capacitance transient spectroscopy*

Information about the defects in typical Schottky diodes can be obtained by studying the kinetics of the system in recovering its equilibrium after a perturbation due to an application of a voltage or a light pulse. This can be done by measuring the capacitance of the diode as a function of time to determine its variation, which relates to the emission of the captured charge carriers in the depletion region. Here, the capacitance is analyzed using the time t as a variable and not the frequency f nor the DC applied voltage V as in the admittance or impedance investigations. The technique is therefore called the *capacitance transient spectroscopy* since the capacitance is recorded as a function of time.

In order to explain the principle of the technique, let us consider a Schottky contact on an N-type semiconductor and assume one type of trap. The width of the depletion is W_0, in which the trapping centers localized above the Fermi level are empty and the centers localized below the Fermi levels are occupied by electrons (Figure 4.27).

When the voltage pulse is applied to the semiconductor, the width of the depletion region is reduced to W_F. When it is set to zero, the reverse bias increases the width of the depletion region to W_R. In this depletion region, the carriers occupying the trap levels just below the Fermi level before the pulse application are now situated above the Fermi level and can be thermally emitted to the conduction band, emptying the trap centers. The emitted carriers in the space charge region of the conduction band are then swept away by the applied electric field. The capacitance of the Schottky contact increases or decreases when the width of the depletion decreases or increases. Thus, it varies with the time t during the application cycle of the voltage pulse. The junction capacitance transient is characterized by a time constant τ, which is equal to the reciprocal emission rate of the charge carrier (e_n in this example) in the trap level. The time dependence of the capacitance can be expressed by:

$$C(t) = C_0 + \Delta C \exp\left(-\frac{t}{\tau}\right) \qquad [4.102]$$

where C_0 is the capacitance of the junction at equilibrium (without applied voltage pulse), and ΔC is the capacitance variation during the charging time, that is, between the beginning and the end of the pulse.

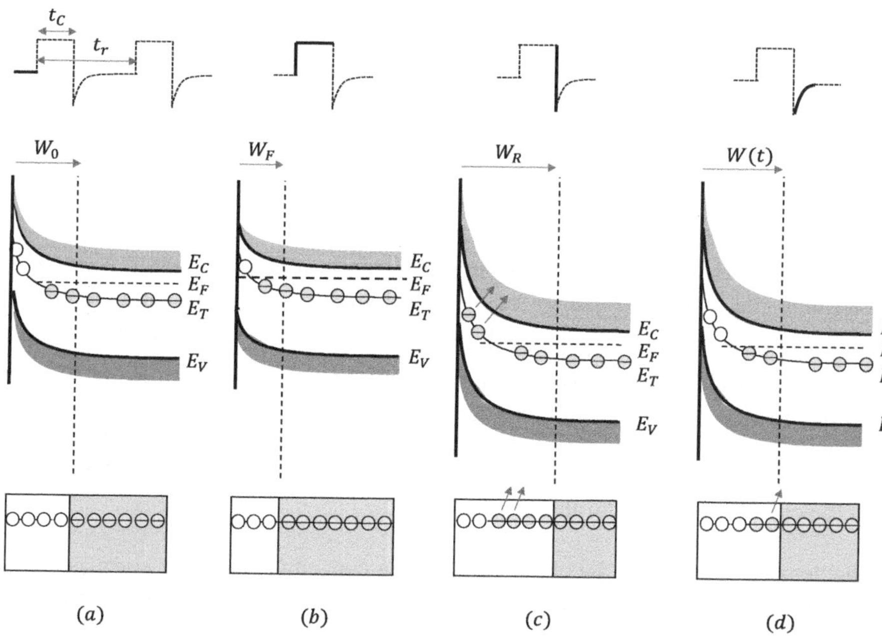

Figure 4.27. *Variation of the space charge region with an applied voltage pulse of charging time t_C and repetition time t_r: (a) before the applied pulse, (b) during the applied pulse, (c) end of the applied pulse and (d) during the charge emission from trapping centers. At the end of the charge emission, the space charge region goes back to its initial state. For a color version of this figure, see www.iste.co.uk/nguyen/defects.zip*

The occupancy of the trapping centers by the charge carriers during the application of the pulse voltage is schematically shown in Figure 4.28.

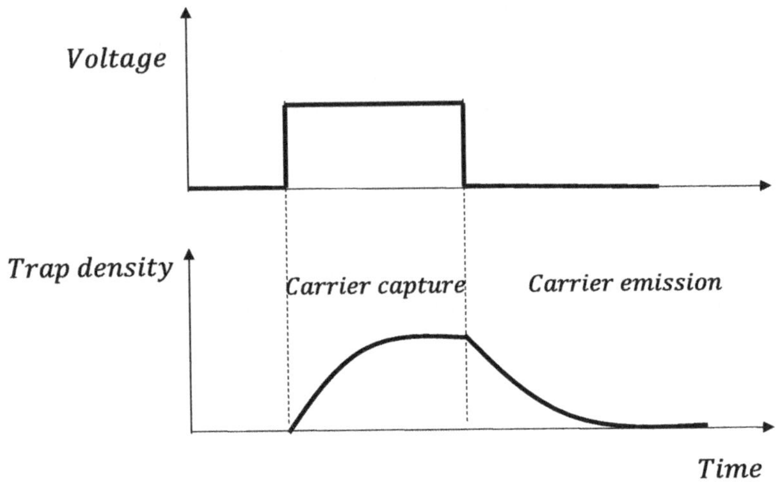

Figure 4.28. *Schematic representation of the occupancy of trap states in the depletion region when a pulse voltage is applied to the Schottky contact*

By recording the capacitance transient as a function of temperature, it is possible to determine the activation energy of the trap emission using equation [3.39] for electron traps: $e_n = \Gamma T^2 \exp\left(-\frac{E_{nA}}{kT}\right)$. The slope of the plot $\ln(\Gamma T^2)$ versus $1{,}000/T$ will provide the electron emission activation energy E_{nA}. Knowing the trap activation energy, the capture cross-section for electron or hole traps can be determined by using equations [3.39a] or [3.39b]. It should be noted that the frequency of the pulse should be chosen to be sufficiently slow in order to stay below the cut-off frequency of the material bulk.

The density of traps can be obtained from the capacitance change corresponding to the complete filling the trap with the applied pulse. The amplitude of the transient is linearly proportional to the density of traps N_T by:

$$N_T = 2\frac{\Delta C}{C}(N_D - N_A) \qquad [4.103]$$

where ΔC is the capacitance variation at the end of the pulse application, C is the capacitance of the diode before the pulse application, and $(N_D - N_A)$ represents the net charge carrier concentration in the depletion region where the traps are measured.

As an example, the transient capacitance of an $ITO/PPV/Al$ diode is shown in Figure 4.29 (Campbell et al. 2000a). The increase in capacitance with time is very long, suggesting that the defect centers are deep, and the transient is not truly an exponential function of time, indicating that the defect centers are distributed in energy. Using a Gaussian function for modeling the decay function $f(t)$ of the capacitance transient, the fit of the experimental data yields a density of traps of $\sim 5 \times 10^{17}\, cm^{-3}$ and a trap depth of $\sim$ 0.75 eV. The defects are related to the presence of air or water in the polymer.

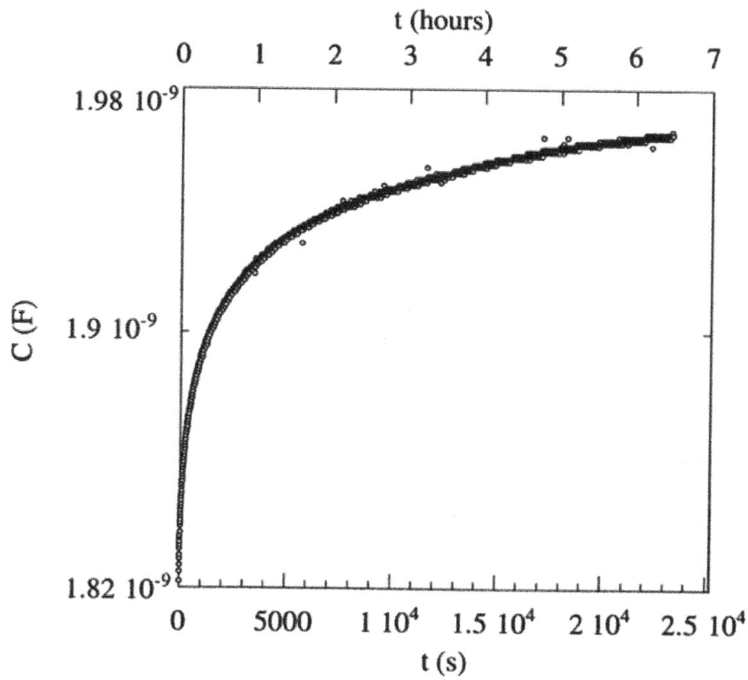

Figure 4.29. *Transient capacitance $C(t)$ of an $ITO/PPV/Al$ structure (from Campbell et al. 2000, p. 273)*

4.3.4.2. *Deep-level transient spectroscopy (DLTS)*

The DLTS is based on the capacitance transient spectroscopy by introducing the rate window concept that makes it possible to directly record the transient spectrum of the device at a given temperature. It consists of measuring the change in occupancy of a given trapping center, through the capacitance change ΔC, at two instants $t = t_1$ and $t = t_2$ during the carrier emission process. The capacitance change is then $\Delta C = C(t_1) - C(t_2)$. We can then write:

$$\Delta C = C(t_1) - C(t_2) \propto \left[\exp\left(-\frac{t_1}{\tau}\right) - \exp\left(-\frac{t_2}{\tau}\right)\right] \quad [4.104]$$

The normalized DLTS signal is defined as:

$$S(T) = \frac{[C(t_1) - C(t_2)]}{\Delta C(0)} = \left[\exp\left(-\frac{t_1}{\tau}\right) - \exp\left(-\frac{t_2}{\tau}\right)\right] \quad [4.105]$$

$\Delta C(0)$ is the capacitance due to the pulse at $t = 0$.

Differentiating $S(T)$ with respect to τ and setting the results equal to zero, we can determine the maximum change in capacitance, which occurs when $\tau = \tau_m$ with:

$$\tau_m = \frac{t_2 - t_1}{\ln(t_2/t_1)} \quad [4.106]$$

τ_m is called the time window.

In the original method proposed by Lang (Lang et al. 1974) (also called the *boxcar method*), the rate window is implemented by a double boxcar integrator, which produces an output proportional to the difference of input signal at instants $t = t_1$ and $t = t_2$. These instants are pre-determined so that the measurements will be made with the same rate window, that is, the output signal $C(t_1) - C(t_2)$ is measured at two precise instants after the filling pulse. The time window τ_m is a known preset time constant in this method.

When the temperature of the junction varies, the capacitance transient varies exponentially with a time constant τ according to equation [4.102]. The output signal of the boxcar representing the capacitance variation ΔC will be at a maximum when the time constant τ of the exponential decay is equal to the time window τ_m. The emission rate of carriers from the trapping centers can be then determined by $\tau_m = \tau = \left(e_{n,p}\right)^{-1}$.

The DLTS spectrum is the plot of the capacitance transient through the defined rate window as a function of temperature. The presence of a defect level in the depletion region is indicated by a capacitance peak in the spectrum corresponding to a temperature T_{max}. An example of DLTS spectrum is given in Figure 4.30. The structure of the studied device is $ZnO/CdS/Cu(In,Ga)Se_2$, and the measurements are performed on two samples, which show principally hole traps of activation energy of $0.26\ eV$. The second DLTS peak at high temperature in sample #2 is assigned to deep traps located at midgap but could not be resolved to provide trap parameters (Igalson and Zabierowsk 2000).

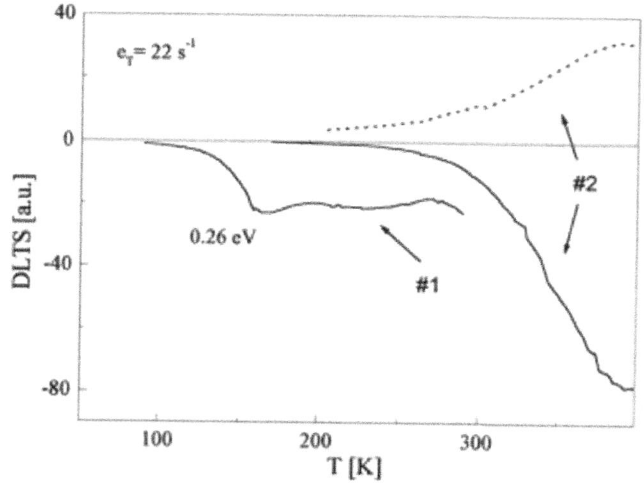

Figure 4.30. *DLTS spectra for two samples: #1 and #2 of CIGS of efficiencies respectively 11% and 14% (from Igalson and Zabierowsk 2000, p. 371)*

4.3.4.3. *Variations of the deep-level transient spectroscopy (DLTS)*

In inorganic semiconductors, the carrier mobility is high and facilitates the transport processes including trapping and detrapping from trap states. The DLTS technique introduced by Lang is therefore widely used for the investigation of defects in devices using conventional semiconductors, without noticeable difficulties. In contrast, in organic materials, as the carrier mobility is low, trapping and emission of carriers by and from trapping centers take more time to be accomplished. In the DLTS technique, the carrier trapping is produced with the application of a voltage pulse of a duration equal to the charging time t_C, which should be long enough to make it possible to fill most of the available trap states. Furthermore, it should exceed, by far, the relaxation time of carriers to allow the system to reach its equilibrium state after the voltage pulse application. In other words, the pulse repetition time t_r should be large enough such that the transient has sufficient time to go back to its initial value before the next pulse is applied. If these conditions are not fulfilled, the DLTS spectra analysis will be affected, and the determination of the trap parameters would not be accurate.

A second factor that may influence the accuracy of the capacitance transient measurement is a high series resistance of the layer interface. A high circuit impedance will cause a large time constant to be measured, which is due to the reactance–capacitance combination of the set-up and the device and not to the deep defects in the semiconductor. In fact, when the series resistance R_S is high, by using the equivalent circuit of Figure 4.18, the measured transient capacitance C_m can be expressed as a function of true capacitance C of the depletion region as:

$$C_m = C[1 + (\omega C R_S)^2] \quad [4.107]$$

In this case, the amplitude of the transient is significantly higher than in the case of a low series resistance ($R_S{\sim}0$) and the measured $\Delta C_m/C_m$ ratio will be reduced leading to a low value of the trap density given by equation [4.103], which should be corrected after detection (Broniatowski et al. 1983).

On the contrary, as degradation processes of organic materials are sensitive to the temperature, it is not convenient to perform DLTS measurements in the high temperature range because of the risk of failure to the devices. As a result, it is relatively difficult to detect very deep traps in the band gap, which need a high energy supply to release the captured charge carriers.

Therefore, not only the measurement protocol should be defined with care but also the quality of the measurement instruments should be adapted to the semiconductors to be characterized. This is particularly true when using the DLTS technique to determine trap parameters in organic semiconductors. Therefore, several variations of the technique have been tentatively tried for organic devices with more or less success.

We can distinguish between techniques that look for improving the analysis and treatments of the measurement results and those that look for improving the measurement itself. Here, we will consider some principal DLTS variations that are proven and currently used and those that can be efficiently applied to organic semiconductors and devices.

4.3.4.3.1. Laplace DLTS

Because of the inaccuracy in the trap parameter determination when analyzing the DLTS spectra, especially for those traps having sufficiently close emission rates, a method referred as Laplace DLTS or LDLTS is proposed to improve the resolution of the spectrum and to discriminate between energetically close trap levels. The mathematical method is based on the use of the reverse Laplace transform of a spectral function in order to fit the experimental DLTS spectrum with exponential functions (Dobaczewski et al. 1994).

Let $f(s)$ be the recorded DLTS transient spectrum of emission rates, and $F(s)$ be the spectral density function. We have:

$$f(t) = \int_0^{\infty} F(s) e^{-st}\, ds \quad [4.108]$$

where s is the exponential decay rate. In this equation, the capacitance transient is the Laplace transform of the true function spectral density function. As the capacitance transient is supposed to be an exponential function of time, $F(s)$ is a delta function located at $s = e_{n,p}$. Therefore, by the inverse Laplace transform, the transient capacitance will be represented by a sum of constituent exponentials; each of them is a Dirac delta function. The DLTS peak associated with defects is generally broad because of the superimposed time constants from different defect states, whereas the resolved Laplace DLTS peak for the same defects are narrow separated peaks. As a result, the emission rate of the defects is, of course, more accurately determined by the Laplace DLTS technique, and the separation of closely spaced defect levels is possibly achieved. The plot of $F(s)$ as a function of the emission rate will provide a spectral function which can be exploited to provide parameters of the defect states and information on the physical processes, which are associated with the studied defects (Figure 4.31).

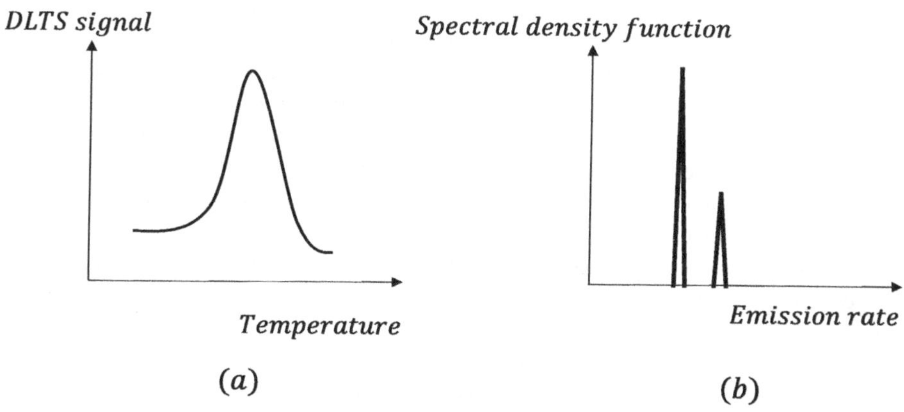

Figure 4.31. *Schematic representation of a measured DLTS spectrum and its corresponding spectral density function obtained by the reverse Laplace transform*

It should be noted that in order to correctly separate transients, a high order of signal-to-noise of the recorded signal is required, and a high number of transients should be averaged. The DLTS spectrum should be recorded at a fixed temperature (isothermal condition) and a mathematical algorithm, which performs the reverse Laplace transform, is applied to extract the emission rate spectrum shown in Figure 4.31b.

A detailed description of the Laplace DLTS and their applications to study the deep-level defects in numerous conventional semiconductors, including alloy systems, are reviewed by Dobaczewski et al. (2004).

4.3.4.3.2. Other variations of the conventional DLTS

For most of the conventional semiconductors, which have a high density of free carriers (electrons or holes) as compared to the defect density, the DLTS technique is successfully applied to determine the defect parameters since the capacitance transient signal can be conveniently detected, giving reliable information.

In materials of low free carrier density (lower than the defect density), the intensity of the capacitance transient signal is weak and is more difficult to be measured. The measurements of such signals are much less reliable because they can be buried in the electrical noise. Therefore, several approaches have been proposed to use the rate window concept of the DLTS technique and to modify either the excitation source or the detection of transient signals.

For instance, DLOS (Deep-Level Optical Spectroscopy) (Chantre et al. 1981) is a currently used technique for optical excitation to enable the trapping of charge carriers in the depletion region of a Schottky barrier or of a P–N junction. It is particularly efficient for detection of deep-level defects in low-conductivity materials such as organic semiconductors, in which the thermal emission techniques like DLTS may not be sufficiently accurate for determining the defect parameters. The technique is based on the trap filing by exposing the device to monochromatic light pulses to change the occupancy of the deep-level defects, and to measure the relevant changes in capacitance as performed in DLTS. There are several DLOS modes of operations, which depend on the time dependence of the excitation and the changes in capacitance. The light or electrical excitation can be used continuously or by pulses. Among these variations, the techniques such as *photo-induced current transient spectroscopy* (PICTS) (Matthiew 2003) or *photo-deep-level transient spectroscopy* (P-DLTS) (Mooney 1983) are usually applied to semi-insulating materials, and do not require the formation of a depletion region.

Other alternative techniques to the capacitance DLTS concern the choice of the transient signals to be detected and analyzed. In these techniques, the electrical excitation is applied to the device in the same manner as in conventional DLTS, but the detected transient signals are different. We can distinguish between the voltage transient (*voltage-based transient spectroscopy* or V-DLTS) (Goto et al. 1973), the current transient (*current-based transient spectroscopy* or I-DLTS (Borsuk and Swanson 1980)) and the charge transient (*charge-based transient spectroscopy* or Q-DLTS (Kirov and Radev 1981)). For these techniques, it is assumed that the defect density is small as compared to that of the donor or acceptors $\left(N_T \ll N_{D,}N_A\right)$, the voltage pulse V is applied to the device to inject charge carriers and fill the trap states. The depletion region is widened at the end when the pulse goes back to its steady state by emission of the trapped charge carriers and the transient signal is measured. As in the case of capacitance DLTS, the decay of the transient signal is related to the emission of the trapped carriers and the process is supposed to be thermally activated of activation energy. The emission rate of carriers is given by equation [3.39] for trapped electrons.

In the V-DLTS technique, the voltage transient is recorded under *constant capacitance conditions* by using a feedback loop to adjust to bias voltage V and to keep the depletion width constant (Johnson and Bartelink 1979). The change in bias V is proportional to the change in charge of the depletion region. The carrier emission from the defect states is analyzed by the voltage variation with time and the V-DLTS signals plotted as a function of temperature show emission peaks corresponding to defect levels. Recording the temperature of each emission peak for different emission rate windows provides the trap parameters analyzed by the technique used to study deep trap states in a similar way to the capacitance DLTS. The voltage transient for electron emission is (Blood and Orton 1992):

$$\Delta V(t) = -\Delta V_0 \, exp[-e_n(t)] = -V(\infty)\frac{N_t}{N_D} exp[-e_n(t)] \qquad [4.109]$$

where ΔV_0 is the transient amplitude, N_t is the trap density, N_D is the donor density and $V(\infty)$ is the voltage at steady state.

In the I-DLTS technique, the charge released from trap states is detected by a current flowing in the external circuit. The current transient is given by (Neugebauer et al. 2012):

$$I(t) \propto \int_0^{\infty} N_t(e_t) e_t \exp(-te_t) de_t \qquad [4.110]$$

where e_t is the trap emission rate, and $N_t(e_t)$ is the distribution of emission rates corresponding to a distribution of trap states $N_t(E)$. The I-DLTS spectrum is the plot of the current transient as a function of temperature and shows current peaks, representing the trap emission rates, which correspond to trap levels analyzed using the technique used for the conventional DLTS (Wessels 1976).

4.3.4.3.3. Charge-based DLTS or Q-DLTS

As previously mentioned, most of the DLTS variations have been successfully applied to conventional semiconductors for accurate measurements of defect parameters even in materials with high or low trap density, and with shallow or deep traps. When using these techniques for organic semiconductors, the obtained results are much less reliable. One of the encountered difficulties of the measurement is the low mobility of the charge carriers, which influences the transport process, and consequently, the measurement protocol. Using the same experimental set-up and measurement parameters for conventional semiconductors with high charge carrier mobility would lead to inaccurate results for trap determination.

Here, we introduce the charge-based DLTS technique or Q-DLTS, which has proven to be efficient in the study of defects in organic materials and devices. This technique is based on the detection of the charge transient resulted from the carrier emission from the trapping centers after application of a voltage pulse. It was developed to study defects in Metal–Insulator–Metal (MIS) structure (Kirov and Radev 1981) and diodes (Farmer et al. 1981). The main advantages of the technique over the conventional DLTS are: (i) the simplicity of the measuring set-up, (ii) the wide range of the relaxation times that can be covered and (iii) the high sensitivity especially for deep levels with low emission rates. Contrarily to the capacitance detection ΔC, which decreases when the ratio $C(t)/C_0$ decreases, the charge detection ΔQ tends to its maximum when the thickness of the insulator decreases or when the resistivity of the substrate increases. Therefore, this technique is suitable for studying defects in thin insulator layers or semiconductors of high resistivity.

The technique is briefly described as follows. At a given temperature T, a voltage pulse ΔV is applied to the device for a charging time t_C to fill the trap in the vicinity of the Fermi level E_F, that is, the regions where the Fermi level crosses a trap level. The pulse is then set to zero, and the trapped charges are released from trapping centers as the samples returns to its equilibrium states. A transient current is generated in the external circuit and is integrated to give the quantity of the released charge ΔQ at two successive times t_1 and t_2: $\Delta Q = Q(t_2) - Q(t_1)$. The released charge ΔQ is measured for increasing values of the time window $\tau = \frac{t_2-t_1}{\ln(t_2/t_1)}$ (equation [4.106]).

The Q-DLTS spectrum is the plot of the released charge ΔQ as a function of time window τ. Assuming that the charge released from a single trap level of energy E_t from the band edge is an exponential function of time, ΔQ is given by:

$$Q(t) = Q_0\left[1 - \exp\left(-\frac{t}{\tau_{rel}}\right)\right] \qquad [4.111]$$

where Q_0 is the amount of charge trapped during the filling pulse ΔV of duration t_C, and τ_{rel} is the relaxation time of the trap.

The Q-DLTS signal can be written as:

$$\Delta Q = Q_0\left[\exp\left(-e_{p,n}t_1\right) - \exp\left(-e_{p,n}t_2\right)\right] \qquad [4.112]$$

From equation [3.39], the hole (electron) emission rate $e_{p,n}$ is given by:

$$e_{p,n} = \tau_{rel}^{-1} = \sigma_{p,n}\Gamma\, T^2 \exp\left(-\frac{E_t}{kT}\right) \qquad [4.113]$$

where Γ is a constant given by:

$$\Gamma = 2 \times 3^{1/2}\left(\frac{2\pi}{h^2}\right)^{3/2} k^2 m_{p,n}^* \qquad [4.114]$$

where $m_{p,n}^*$ is the hole (electron) effective mass.

Usually, the ratio t_2/t_1 is experimentally chosen to be a constant $t_2/t_1 = \alpha$. It can be shown that the Q-DLTS spectrum goes through a maximum ΔQ_{max} when the emission rate matches the rate window, that is, when $e_{p,n} = \tau^{-1} = \tau_m^{-1}$ (Figure 4.32). Accordingly, by recording the Q-DLTS spectrum at several temperatures, we obtain a set of trap relaxation times τ_m and a plot of $ln(\tau_m^{-1}T^{-2})$ versus $1{,}000/T$ provides the activation energy E_t and the trap capture cross-section $\sigma_{p,n}$ by the slope and the intercept with the y-axis respectively. From equations [4.113] and [4.114], it can be seen that only the intercept of the plot with the y-axis, that is, $\sigma_{p,n}$, depends on the carrier effective mass while the slope, that is, E_t, is not. In practice, $m_{p,n}^*$ is taken to be the electron rest mass $m_{p,n}^* = m_0$. The maximum of the Q-DLTS spectrum is given by (Arora et al. 1993):

$$\Delta Q_{max} = Q_0\left[\alpha^{1/(1-\alpha)} - \alpha^{\alpha/(1-\alpha)}\right] \qquad [4.115]$$

In practice, α is usually set equal to 2 so that $\Delta Q_{max} = Q_0/4$. Under saturating filling pulse conditions, the trap density can be obtained from:

$$N_T = 4\Delta Q_{max}/|e|Ad \qquad [4.116]$$

where A is the area of the sample and d is its thickness.

From equation [4.112] and by using $t_2/t_1 = 2$, the charge variation ΔQ as a function of time window is given by:

$$\Delta Q(t) = Q_0\left[\exp\left(-\frac{\tau}{\tau_m}\ln(2)\right) - \exp\left(-\frac{\tau}{\tau_m} \times 2\ln(2)\right)\right] \qquad [4.117]$$

This equation will be used to build the spectrum of a Q-DLTS component when its rate window can be experimentally determined.

As an example, Figure 4.33 shows the Q-DLTS spectra recorded in MEH-PPV-based diodes at different temperatures and the plot of $ln(\tau_m^{-1}T^2)$ versus $1/T$, which allows for the determination of the trap activation energy and the capture cross-section (Nguyen et al. 2006). We observe that with increasing temperature of the sample, the spectra increase in intensity and the time window decreases indicating that more carriers at a deeper level are released.

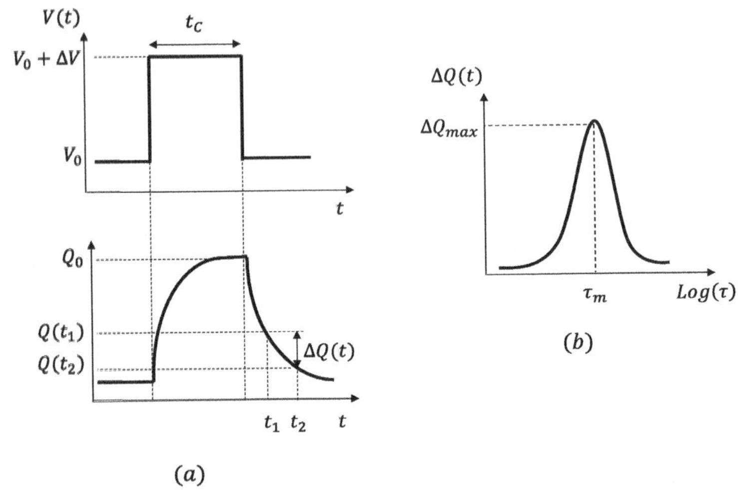

Figure 4.32. *Schematic diagram of a measured Q-DLTS spectrum: (a) timing diagram showing the voltage pulse and the charge variation in and from the traps during one cycle of the measurement and (b) the measured Q-DLTS spectrum at a given temperature T. In figure (a), V_0 is an off-set voltage that is generally set to zero. It can be set to a specific value in some situations (study of defects in a defined spatial region, for example)*

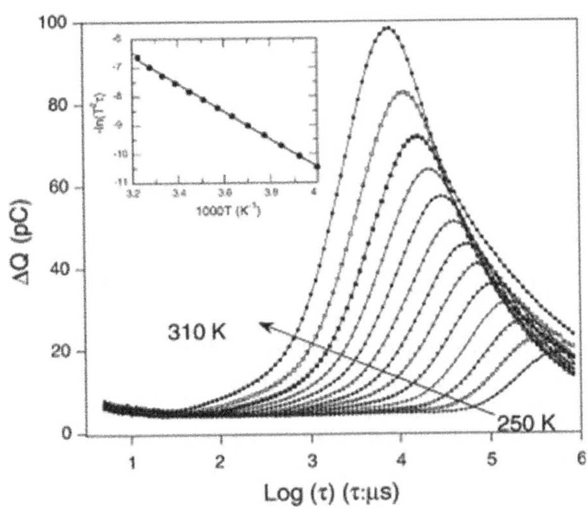

Figure 4.33. *Q-DLTS spectra obtained from an $ITO/MEH\text{-}PPV/Al$ device using a charging voltage $V_0 = 0\,V$, a voltage pulse $\Delta V = 10\,V$, and a charging time $t_C = 1\,s$ for different temperatures within the range 250–310 K. Inset: Arrhenius plot (from Nguyen et al. 2006, p. 338)*

Like other variations of the DLTS technique, the expressions used to determine the trap parameters from the Q-DLTS measurements are accurately valid when all the traps are filled during the experiments. In particular, for organic semiconductors, the charging time t_C should be appropriately set to exceed the trap relaxation time to avoid an incomplete filling of traps. The influence of the finite duration of the charging pulse on the accuracy of trap parameters measured by the Q-DLTS technique is investigated by Barancok et al. (1999). They show that reducing t_C leads to a decrease of the Q-DLTS peak intensity for both discrete and distributed traps. Furthermore, for discrete traps, the peaks are shifted to higher temperatures while their positions are unchanged for distributed traps. While for inorganic semiconductors, the charging times used for saturating pulses are rather short, in the range from 1 to 100 *ms*, they are much longer for organic semiconductors. In PPV devices, the charging time for filling traps is $t_C \sim 1\,s$ (Gaudin et al. 2001). To determine the correct t_C to be used for the measurements, the Q-DLTS spectra are recorded at a fixed temperature by using increasing charging times of the experimental system. The variation of the maximum value ΔQ_{max} of the spectrum is then plotted as a function of t_C (Figure 4.34), which shows a saturation in the number of traps filled and gives the value of the charging times to be chosen. Depending on the material and the device used, it may happen that the charge saturation is not achieved. In this case, the traps are not completely filled, and their density cannot be accurately determined.

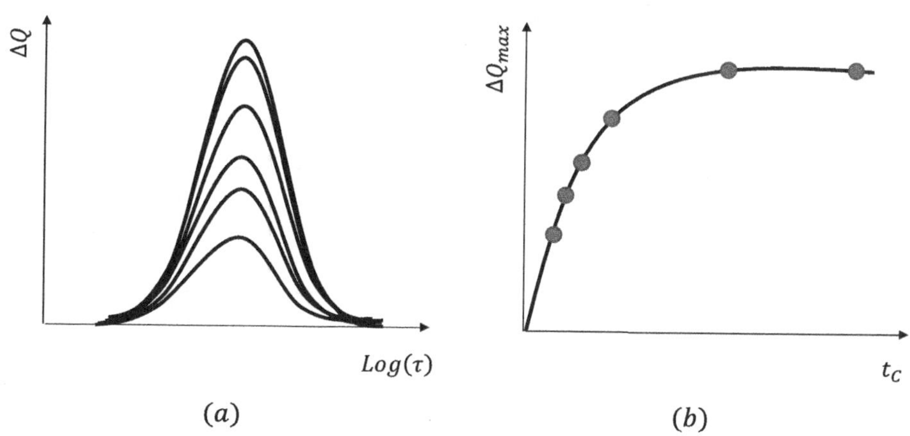

Figure 4.34. *Schematic representation of the effect of charging time to fill traps: (a) variation of Q-DLTS spectrum with charging time and (b) variation of the maximum value ΔQ_{max} of the spectrum with charging time. For a color version of this figure, see www.iste.co.uk/nguyen/defects.zip*

It should be noted that to fully exploit the Q-DLTS results, the spectra are recorded for a large range of temperatures, and for each temperature, they are recorded by using different charging times, in the range from 1 millisecond to a few 10 seconds. If the use of long charging times makes it possible to ensure the complete charge release from deep traps, short charging times are useful to identify low-density or shallower traps. Indeed, when several trap levels are present in the band gap (as in the case of the distributed traps), the Q-DLTS intensity of the low-density traps is gradually hidden by that of the high-density traps when the charging time increases. However, these low-density traps can be observed if they can release the trapped carriers by a short voltage pulse, that is, if they are shallow traps or if their capture cross-section allows the charge emission. The identification of a Q-DLTS peak, that is, its rate window, is necessary to resolve the spectrum into trap components, each of them corresponds to an energy level, which composes the density of defect states. For the resolution of the spectrum, equation [3.179] is applied to each identified rate window.

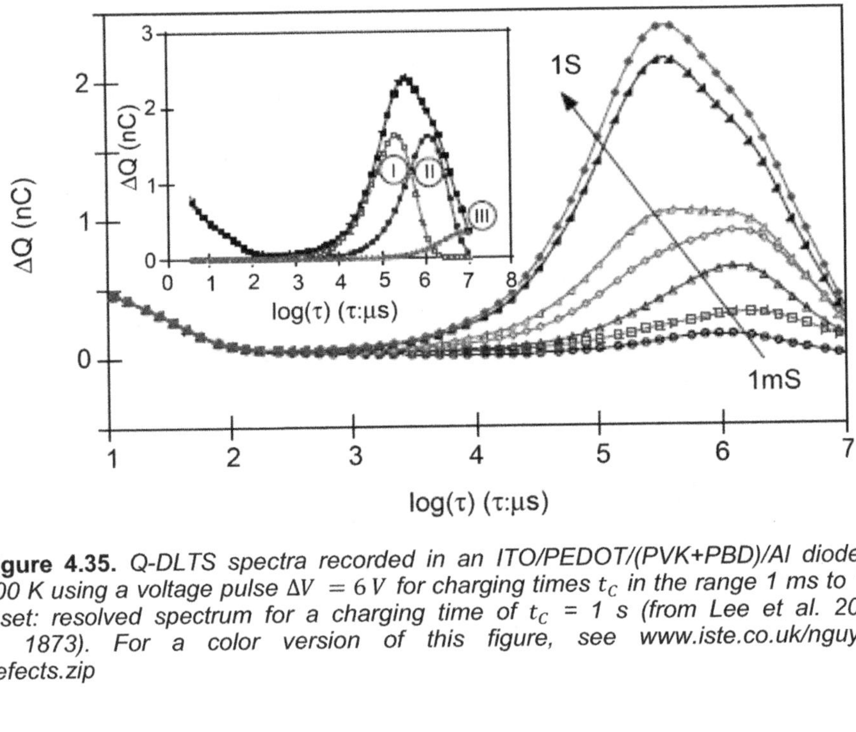

Figure 4.35. *Q-DLTS spectra recorded in an ITO/PEDOT/(PVK+PBD)/Al diode at 300 K using a voltage pulse* $\Delta V = 6\,V$ *for charging times* t_C *in the range 1 ms to 1 s. Inset: resolved spectrum for a charging time of* t_C *= 1 s (from Lee et al. 2010, p. 1873). For a color version of this figure, see www.iste.co.uk/nguyen/defects.zip*

As an example, Figure 4.35 shows the Q-DLTS spectra of a (PVK+PBD)-based diodes recorded at room temperature with different charging times t_C from 1 ms to 1 s (Lee et al. 2010). Two apparent charge peaks referred to as I and II are observed. Peak II appears first in the spectrum when short t_C are used, then peak I starts growing and becomes dominant with long t_C. Peak III is an uncomplete peak and could not be analyzed for the range of t_C used in this experiment. The positions of the peaks (τ_m) are then used to compute the Q-DLTS components by equation [4.117] as shown in the inset.

Q-DLTS measurements allow for determination of trap parameters such as activation energies, capture cross-sections and trap concentrations in structures whose capacitance does not depend on the charge state of the traps. The method can therefore be applied for study of defects in organic semiconductors and devices with a good reliability as compared to other techniques.

4.3.4.4. *DLTS measurements of organic semiconductors and devices*

DLTS technique is currently used for characterizing defects in inorganic semiconductors and remains little known in organic electronic field, probably due to the fact that the organic semiconductors are dielectric-like materials and applying the same measurement protocol and experimental conditions as for studying defects in crystalline materials would not provide expecting results. The technique and its variations have been gradually applied with success in conjugated polymers and small molecules and have then become increasingly used for characterizing defects in organic materials. Typical applications of the technique are given for conjugated polymers, small molecules and hybrid materials used in organic devices.

Jones et al. (1997) use capacitance DLTS to investigate traps in P3MeT Schottky diodes of structure $Au/P3MeT/Au$. They obtain a DLTS spectrum in the temperature range from 20 to 60°C, proving that the technique could be applied to organic diodes but give no trap parameters.

Later, trap investigations in PPV by capacitance transient measurements are performed in $ITO/PPV/Al$ diodes in the temperature range from 240 to 320 K (Campbell et al. 2000b). The variation of relative capacitance $\Delta C/C$ is plotted as a function of time for various temperatures, and then fitted by a model using a Gaussian distribution of transport states. The trap parameters are determined to be $E_t = 0.75\ eV$ for the activation energy and $N_t{\sim}4 \times 10^{17}\ cm^{-3}$. The same polymer in diodes of configuration $ITO/PPV/MgAg$ is studied by Q-DLTS measurements carried out in the range 100–315 K (Gaudin et al. 2001). Two types of traps are identified. Type I traps have an activation energy of $E_t{\sim}\ 0.50\ eV$, relative to the HOMO and a capture cross-section of the order of 10^{-18}- $10^{-16}\ cm^2$. They are assigned to acceptor-like hole traps. Type II traps have an activation energy of $E_t{\sim}0.40\ eV$, relative to the LUMO and a capture cross-section of the order of $10^{-19}\ cm^2$. They are assigned to donor-like electron traps. The density of both traps is of the order of $N_t{\sim}10^{17}\ cm^{-3}$.

The electronic levels in MEH-PPV are studied by capacitance transient $\Delta C/C$ measurements of diodes of structure $Si/MEH\text{-}PPV/Au$ in the temperature range from 100 to 300 K (Stallinga et al. 2000). By fitting the experimental curves by the expression of the emission rate (similar to equation [3.39] for electrons), four trap levels are found. Two of them are electron traps of activation energy of 0.48 and 1.3 eV. Two other levels are hole traps of activation energy of 0.30 and 1.0 eV. Q-DLTS is used for trap measurements in diodes of structure $ITO/MEH\text{-}PPV/Al$ in the temperature range from 100 to 310 K (Nguyen et al. 2006). Only one distributed trap level can be completely determined and is assigned to hole traps with an activation energy of 0.3–0.4 eV, a capture cross-section of 10^{-20}– 10^{-18} cm^2, and a trap density of $N_t \sim 10^{15}$ cm^{-3}. A second trap level is observed by the charge peak with a relaxation time higher than 1 s and could not be fully determined by the instrument and the experimental temperature limitations.

We note that the results of trap measurements in both PPV and MEH-PPV differ from each other. The techniques used in the investigations are, of course, not at stake; they give reliable results. The trap parameters discrepancy can be explained by the quality of the conjugated polymers, which depends strongly on the synthesis methods.

Example of trap study in small molecules by DLTS technique can be found in pentacene-based diodes of structure $Au/pentacene/Al$ (Juhasz et al. 2015). The transient capacitance response is recorded during the diode discharging at various temperatures from 100 to 400 K. Using the Fourier analysis of the capacitance transients (Weiss and Kassing 1988), the DLTS spectrum is analyzed providing six trap levels of activation energy of $0.25, 0.26, 0.31, 0.32, 0.43$ and 0.46 eV. They are assigned to both chemical and structural defects in pentacene.

Defects in P3HT:PCBM are studied by Q-DLTS technique to study the degradation process of the blend used as an absorber in solar cells (Nguyen et al. 2012). The device structures are $ITO/PEDOT{:}\,PSS/P3HT/CaAl$ and $ITO/PEDOT{:}\,PSS/P3HT{:}\,PCBM/CaAl$, and the spectra are recorded at various temperature from 100 to $310K$. The measurements are performed using different charging times from 1 ms to 10 s to detect the charge peaks, which make it possible to determine the relaxation times τ_m. The spectra are resolved into components using equation [4.117]. There are six complete trap levels in the P3HT diodes whose activation energies are 0.09, 0.20, 0.26, 0.31, 0.36 and 0.47 eV. Their densities are in the range of $3.0 \times 10^{16} - 1.0 \times 10^{17}$ cm^{-3}, and their capture cross-sections vary from 10^{-21} to 10^{-16} cm^2. In P3HT:PCBM diodes, the spectra show the same trap levels with an additional level with a long relaxation time ($> 10\ s$), which is incompletely resolved. This trap level is related to PCBM. Comparing the trap density of P3HT and P3HT:PCBM diodes, it is found that the trap density in P3HT:PCBM is much higher than that of P3HT, and the enhancement in density is not uniform as deep traps are found to be more affected than the shallow ones. This observation suggests that incorporation of PCBM into the P3HT matrix creates a highly disordered structure.

In perovskite-based devices, DLTS has been used for characterizing defects in different perovskite structures. For instance, $CH_3NH_3PbI_3$ perovskites prepared by one step and two-step deposition are studied using diodes of structure $FTO/TiO_2/mesoporous\ TiO_2/Perovskite/Spiro\text{-}OMeTAD/Au$ (Heo et al. 2017). Capacitance DLTS measurements are performed in the temperature range 20–430 K using a pulse width of 10 ms and a pulse period of 10 ms. Two defect levels are found with activation energy of 0.62 and 0.76 eV for both types of perovskite. The capture cross-sections are in the range of 2.5×10^{-15} to $5 \times 10^{-14}\ cm^2$. The trap density of the one-step perovskite at the 0.62 eV level is higher than the 0.76 eV level ($1.3 \times 10^{15}\ cm^{-3}$ and $3.9 \times 10^{14}\ cm^{-3}$). For the two-step perovskite, the trap density of the 0.76 eV trap level is higher than that of the 0.62 eV ($9.5 \times 10^{14}\ cm^{-3}$ and $5.0 \times 10^{14}\ cm^{-3}$). The efficiency of the studied solar cells is related to the presence of defects in perovskite and as the two-step prepared materials containing lower density of defects, its efficiency is higher than that of solar cells using the one-step perovskite as an absorber.

Q-DLTS experiments are performed in a series of devices using $CH_3NH_3PbI_3$ perovskite as an absorber and spiro-OMeTAD, P3HT and PBTTTV-h as the HTL to determine the influence of interfacial traps on the solar cell efficiency (Hsieh et al. 2018). The structure of the devices is *FTO/*TiO_2*/perovskite/HTL/Au*, and the measurement protocol is similar to that used for P3HT:PCBM diodes. By comparing the trap parameters obtained in diodes with different HTL materials, the defects related to each active layer are identified. For the $CH_3NH_3PbI_3$ perovskite, four trap levels are determined with activation energy of $0.36, 0.42, 0.47$ and $0.52\ eV$. Their capture cross-sections are of the order of $10^{-19}\ cm^2$, and their density is higher than $1.6 \times 10^{17}\ cm^{-3}$ (incomplete peaks). On the contrary, defect parameters of each layer (TiO_2 and HTL) could also be determined. It is found that spiro-OMeTAD and PBTTTV-h-based devices have shallow traps favoring charge trapping and reducing charge recombination, therefore improving the charge collection and the cell efficiency.

Defects in crystals $CH_3NH_3PbBr_3$ perovskites are studied by Laplace current DLTS using devices of structure *carbon/perovskite/Au* (Rosenberg et al. 2017). Laplace spectra are derived from current transients measured in the temperature range of 275–300 K, using a pulse width of 1 ms. Two trap levels are determined from the analysis of the spectra: one trap with an activation energy of 0.167 eV, a capture cross-section of $4 \times 10^{-20}\ cm^2$ and a density of $\sim 10^8\ cm^{-3}$, and one trap with an activation energy of 0.204 eV, a capture cross-section of $5 \times 10^{-20}\ cm^2$ and a density of $\sim 10^9\ cm^{-3}$. These defects are all shallow traps in the bulk of the perovskite, which are not recombination centers, and therefore favor the charge dissociation and the efficiency of the devices.

4.3.5. *Time of flight (TOF) and charge carrier extraction by linearly increasing voltage (CELIV) techniques*

These techniques are used for determining the charge carrier dynamics in organic semiconductors, including the lifetime and the mobility of carriers. Through these characteristics, trapping process in devices can be interpreted and understood. However, very few investigations carried out to determine the trap parameters have been reported. Here, a brief description of the methods will be given together with some examples using them in the organic semiconductors and devices.

4.3.5.1. *The TOF technique*

The technique consists of measuring the time t_{tr} taken by the charge carriers generated by a laser pulse near one electrode of a semiconductor sample to reach the opposite electrode at a distance equal to the thickness d of the sample. Excitons are created by the laser in the vicinity of the illuminated electrode and will be dissociated by an applied field. One type of carriers will be extracted to the external circuit and the other carriers will be transported through the sample to the opposite electrode. Let V be the applied voltage to the sample, the charge carrier mobility can be determined by:

$$\mu = \frac{d^2}{t_{tr}V} \qquad [4.118]$$

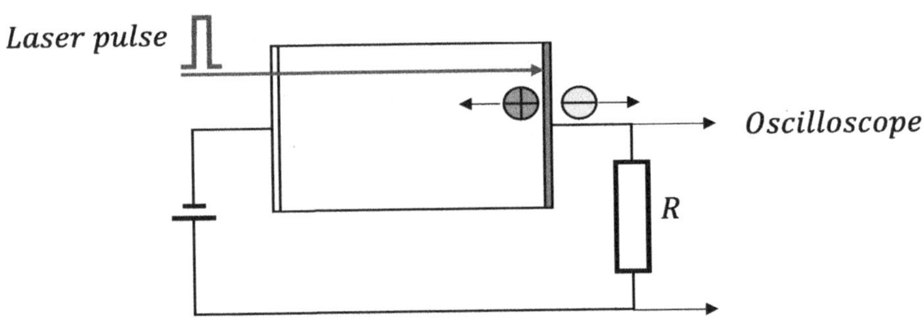

Figure 4.36. *Schematic measurement principle of the time-of-flight experiment. For a color version of this figure, see www.iste.co.uk/nguyen/defects.zip*

Figure 4.36 shows the principle of the TOF technique. The wavelength of the laser pulse is chosen to obtain a maximum absorption of the semiconductor. Several conditions are required for obtaining reliable results for the carrier mobility measurement: (i) the light pulse should be short as compared to the transit time t_{tr}, (ii) the thickness d of the sample should be larger than the absorption depth of the laser to correctly record the charge transient and (iii) the time constant $\tau = RC$ of the circuit should be lower than the transit time, and the dielectric relaxation time $\tau_d = \rho\varepsilon_0\varepsilon_r$ should be larger than t_{tr}, ρ being the conductivity in the dark of the semiconductor.

The photocurrent transient $I(t)$ shows a sharp decay at the origin of time followed by a current plateau or a slow decay to its steady-state value (Figure 4.37). The former is observed for non-dispersive transport, that is, the carriers move uniformly with the same velocity in the semiconductors. The latter part can be seen as an effect of trapping centers on the carrier transport, which is of dispersive nature, and can be analyzed by the model proposed by Scher and Montroll (1975). In the dispersive transport, the photocurrent decays monotonically from the origin of time. According to the model, the decay of the current transient follows a power law as:

$$i(t) \propto t^{-(1-\alpha)} \text{ for } t < t_{tr} \qquad [4.119]$$

$$i(t) \propto t^{-(1+\alpha)} \text{ for } t > t_{tr}$$

The exponent α is the dispersivity of the transport and varies in the range $0 \leq t \leq 1$. In a double logarithmic plot of the photocurrent curves $ln(I)$ versus $ln(t)$, the transit time can be determined by the intersection of the two branches, where the change in the slopes of the straight lines occurs.

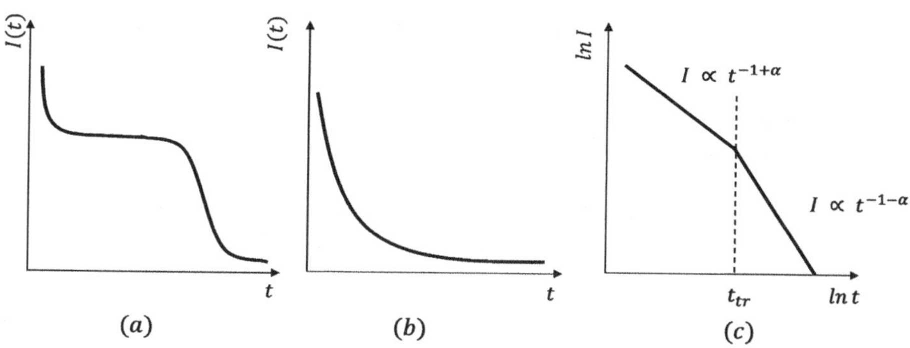

Figure 4.37. *Photocurrent transient: (a) in non-dispersive transport, (b) in dispersive transport and (c) in the double logarithmic plot for dispersive transport*

Initially, the TOF technique is used for determining the charge carrier mobility by applying equation [4.118] to materials in which dispersive transport occurs by the presence of traps or by hopping of carriers (Schwarz 1998). By considering the transport mechanism in disordered materials, that is, hopping of carriers between localized states, the information obtained from the TOF measurements can be used to build the density of states in the band gap (Nagase and Naito 2000). The distribution of localized states is determined by assuming an energetic Gaussian distribution and by applying the GDM Gaussian disorder model (Bässler 1993). The analysis makes use of the Laplace transform (Naito et al. 1994) method to obtain the localized state distribution $g(E)$ from the TOF measurement results. The distribution is then fitted by using the GDM model to extract the parameters such as the energy level E, the energy width of the Gaussian distribution σ, and the attempt-to escape frequency ν.

4.3.5.2. *Charge carrier extraction by linearly increasing voltage (CELIV) technique*

The technique of charge carrier extraction by linearly increasing voltage (CELIV) is used to determine the carrier mobility and density at the same time. The photocarriers are generated in the semiconductor using a light pulse and then, after a delay time t_d, they are extracted by an applied voltage. Its working principle is similar to that of the TOF technique but differs from it by the following aspects: (i) the applied voltage used is a ramp voltage of equation $V(t) = At$, and (ii) the carriers are injected from the electrodes and the photogenerated carriers are uniformly distributed in the bulk of the semiconductor.

The schematic illustration of the CELIV method is shown in Figure 4.38.

The transient current is composed of a plateau, which is related to the geometric capacitive or displacement current j_0 and a CELIV conduction current Δj, which is due to the photogenerated carriers.

Carrier mobility is obtained from the CELIV transient using the continuity of the current and Poisson equation and is given by (Juska et al. 2000):

$$\mu = \frac{2d^2}{3At_{max}^2\left[1+0.36\frac{\Delta j}{j_0}\right]} \qquad [4.120]$$

where d is the distance between electrodes, t_{max} is the time for which the extraction current is maximum $j(t_{max}) = j_{max}$, j_0 is the current from the capacitance, and $\Delta j = \Delta j_{max} = j_{max} - j_0$.

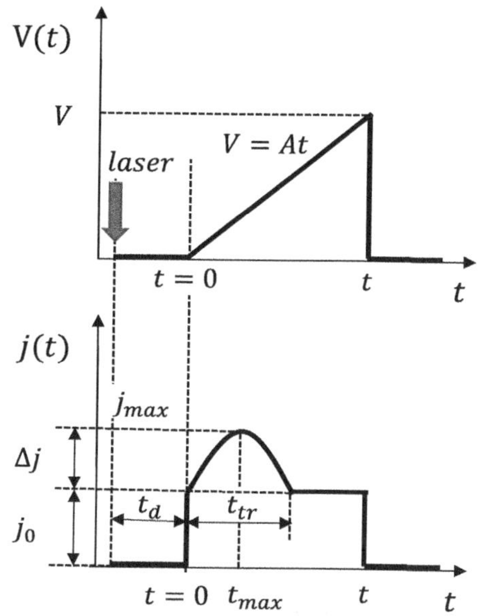

Figure 4.38. *CELIV characteristics of the applied voltage ramp and the current transient obtained in the device. For a color version of this figure, see www.iste.co.uk/nguyen/defects.zip*

Equation [4.120] is often used to determine the carrier mobility of organic thin films because a large thickness is not required for the CELIV technique. However, the accuracy of the results depends on the experimental parameters used for the measurements, especially since the delay time t_d and the trap states can significantly change the shape of the transient (Hanfland et al. 2013) and therefore lead to inaccurate evaluation of the charge carrier mobility. Trap parameters can be determined from CELIV signals assuming that the measured current results from the released of trapped charge carriers under a high filed (Tajima and Yasui 2011). Without going into detail, this approach consists of evaluating the trapped charge Q as a function of the trap energy distribution function $f[\varepsilon, F(t)]$ under a time-dependent electric field $F(t)$, ε being the trap energy level with regard to the band edge of the valence or the conduction band.

The CELIV signal can be expressed as:

$$\frac{J}{A} = \frac{1}{AS}\frac{dQ}{dt} = -\int |e|\,\rho(E)\frac{df(E,F)}{dF}\,dE \qquad [4.121]$$

where A is the rate of voltage increase, S is the area of the sample, and $\rho(E)$ is the trap density. Assuming that the trapped carriers are released with a constant rate

$k(E)$ according to the Poole–Frenkel model, the CELIV signal of the sample at low temperature can be expressed as:

$$\frac{J}{A} = \frac{|e|\beta F^{-\frac{1}{2}}}{2}\rho(\beta F^{-\frac{1}{2}}) \qquad [4.122]$$

where $\beta = \left(\frac{|e|^3}{\pi\varepsilon_0\varepsilon_r}\right)^{1/2}$ is the Poole–Frenkel factor.

From the experimental CELIV signal measured at low temperatures and by using equation [4.122] for fitting the results, the trap density and the trap energy level of the material can be determined.

4.3.5.3. *TOF and CELIV measurements of organic semiconductor and devices*

The TOF and CELIV techniques are widely applied to organic materials for measuring the carrier mobility (Tuladhar et al. 2005), which is an essential parameter to define and to control the performance of semiconductors and devices. Using these techniques, charge carrier trapping in devices can be demonstrated. However, only a few investigations of quantitative trap determination in organic materials have been performed.

For example, bulk heterojunction solar cells using polycarbazole copolymer:fullerene (PCDTBT:PCBM) as an absorber have been studied by TOF and CELIV techniques (Clarke et al. 2012). The recorded TOF transients show a continuous decay without a clear plateau indicating a dispersive transport and charge trapping in the blend. This is further confirmed by the CELIV results, which show a dispersive characteristic with a long tail of charge, suggesting a strong charge carrier trapping. A charge carrier mobility of $\sim 5 \times 10^{-5} cm^2V^{-1}s^{-1}$ is determined with a maximum bimolecular recombination coefficient of $\sim 3 \times 10^{-11} cm^3s^{-1}$, which is close to the Langevin coefficient. However, the trap parameters are not derived from the measurements.

Polycarbonates (PC) doped with molecules of TPD and with PRA have been studied by TOF technique using samples of structure $ITO/doped\ PC/Au$ (Nagase and Naito 2000). Dispersive transport is observed when the concentration of PRA is equal to 1 mol% and the photocurrent transient is analyzed by the GDM model to provide the localized-state distribution in the doped polymer. The energy position of the distribution is $E_P = 0.54\ eV$ with an attempt-to-escape frequency of $1.0 \times 10^{11}\ Hz$.

The CELIV technique has been used to study the trap density function of P3HT:PCBM in the diodes of the structure $ITO/PEDOT{:}PSS/P3HT{:}PCBM/Al$ (Tajima et al. 2012). The current transient is analyzed using the Poole–Frenkel model to provide the trap density function and the carrier drift mobility. The results indicate a trap density of $N_t = 1.0 \times 10^{18} cm^{-3}$, and an energy level of $E_t = 0.087\ eV$. The carrier mobility is estimated to be $\mu \sim 2.3 \times 10^{-6} cm^2 V^{-1} s^{-1}$ at 1.8 K.

In this section, we have namely described some of the most popular and accessible techniques that are frequently used in the organic electronic materials. In the literature, many other variants of these techniques have been proposed and applied to study defects. These variants are seemingly efficient to allow the description and the determination of defects in the systems studied but do not show clear advantages over the original ones. Moreover, for the application of some variants, additional hypotheses or theoretical models are required making the analysis of the measurements more complex.

Other optical or electrical techniques built on working principles, which are different from those previously seen, have been developed and applied to study defects in organic materials. In particular, spectroscopy and microscopy techniques are used to probe and map the defects in organic films. Among these, X-ray-based techniques can provide information on the order and disorder, the crystallinity and the molecular orientation of the material structure. Optical spectroscopy (Raman, Infrared) and microscopy (TEM, AFM, STM) techniques can provide information of the organic film morphology, its local microstructure and its molecular organization in relation with order and defects. Trap densities can also be determined in materials by using some of these techniques. However, the experiments are more often performed to study the organic semiconductor microstructure and to qualitatively characterize defects and disorder in the material. Systematic applications of techniques using optical spectroscopy and microscopy are, unfortunately, still not widely developed (Rivnay et al. 2012).

In the following chapter, we will examine the defects in organic semiconductors in the light of experimental investigations using the techniques described previously.

5

Defect Origins

We have introduced a general description of defects in inorganic and organic semiconductors, and emphasizing the particular structure of conjugated polymers, we expect that the defect concentration in these materials should be much higher than that found in inorganic semiconductors. Indeed, experimental results obtained from electrical measurements generally indicate poor transport with low carrier mobility, and suggest the presence of a significant quantity of defects in the semiconductors. In addition, the lack of stability of organic materials upon exposure to light or air is usually indicative of modifications of their structure by the formation of extrinsic defects. Here, we will distinguish between defects created inside the semiconductors and those found in devices that are composed of several additional layers, and consequently several interfaces. We focus on the experimental aspects and results, and we use this information to get insight into the formation of defects to understand the impact of defects on the operation of devices such as OLEDs, OPVs and OFETs.

5.1. Defects in organic semiconductors

To observe and identify defects in organic semiconductors, spectroscopy and microscopy techniques can be used. The sample is deposited in thin film on an appropriate substrate or in free-standing film. The film thickness is chosen in such a way that only the surface or a part of its volume will be analyzed, and the interface between the film and the substrate will not be.

5.1.1. *Structural defects*

Structural or intrinsic defects are defects that are inherent to the structure of the semiconductor. They include point defects (vacancies, interstitials, disorder) and extended defects (dislocations, grain boundaries). Polymer thin films are expected to have a high density of structural defects due to the high concentration of chain ends in the volume. Thin films of small molecules also have a higher structural defect concentration than single crystal because such films are often polycrystalline, and as a consequence, have a large number of grain boundaries.

As an example of the investigation of structural defects in organic materials, conjugated polymer films of P3HT have been analyzed by IR spectroscopy, using the polarized Brewster angle transmission, whose spectra allow for determining defect features in the film (DeLongchamp et al. 2011). Side-chain disorder is identified by a band centered at $2928\, cm^{-1}$, which is correlated to the conjugation length with the C=C bonds at $1510\, cm^{-1}$ (asymmetric mode from short oligomers) and $1520\, cm^{-1}$ (asymmetric mode from long oligomers). The well-ordered crystalline alkane chains are characterized by bands centered at $2915\, cm^{-1}$ (Macphail et al. 1984).

Other optical techniques have been used to examine the defect state distribution in conjugated polymer thin films of MDMO-PPV through their trapping behavior (Kuik et al. 2011). The *photothermal deflection spectroscopy* (PDS) is a sensitive technique for measuring the optical absorption coefficient of a semiconductor exposed to a monochromatic light beam (Jackson et al. 1981), and allows for the identification of the defects in the sample by analyzing the optical transitions (Goris et al. 2005). The light absorbed by the sample is converted into heat by non-radiative deexcitation, and the heat release will give rise to a change in the refractive index of the sample and that of a fluid in contact with it. Detection of very subtle changes in the optical absorption coefficient is facilitated by a large change in the refractive index of the appropriate fluid, with a small change in temperature. The PDS absorption spectra of two MDMO-PPV samples fabricated by different synthesis methods are recorded and compared. The presence of traps in both samples is observed by a sub-band absorption at about 0.75 eV below the LUMO level with, however, a difference in the absorption intensity, which indicates different defect concentrations in the samples. The defects are hence identified as conformational defects induced in the synthesis of the polymers.

5.1.2. *Impurity defects*

In conventional semiconductors, impurities are intentionally or unintentionally introduced to the structure of the materials during the synthesis process and to the electronic devices during their fabrication. They are considered as chemical impurities because of their chemical nature, which facilitates their incorporation into the crystal lattice. In organic semiconductors, the impurity defects can be formed by the same processes operating in the inorganic counterparts. It is experimentally observed in a range of conjugated polymers that the energy levels of electron traps are located at a rather well-defined region with respect to the vacuum level (Nicolai et al. 2012). This may suggest that these traps are commonly created in conjugated polymers because they have the same origin and are formed by the same type of impurities. These impurities are supposed to be *oxygen* and/or *water*, which are commonly present in the environmental media. They may even be formed in vacuum-evaporated small-molecule thin films and acts as active trapping centers in devices.

The absorption spectrum of $CH_3NH_3PbI_3$ perovskite thin films is measured by the PDS method and shows a high absorption coefficient with a sharp shoulder near its band gap value of ~1.57 eV (De Wolf et al. 2014). Below this shoulder, the variation of the absorption coefficient follows an exponential function over more than four decades with a low *Urbach energy* of 15 meV. This energy is the slope of the exponential part of the absorption curve. When exposed to ambient air with a 30–40% relative humidity, the absorptance between 1.5 and 2.5 eV drops by two orders of magnitude, and the absorption onset shifts from 1.57 to 2.40 eV, suggesting a change in the composition of the films, which is due to moisture ingress, introducing oxygen or water impurities in the material.

Other sources of defects include *chemical impurities*, which can be seen as foreign atoms of different sizes or structures relative to the host atoms or structures. They are introduced by the chemical products used during the synthesis of the organic semiconductors and remain in the material after the synthesis. These products can be produced by solvents, reagents, precursors, residues or by-products from the chemical reactions. They can in turn be oxidized by the contact with residual traces of oxygen and water in the materials, hence increasing the defects in the semiconductor.

Thin films of P3HT obtained from commercial products have been found to contain low concentration of impurities (1,600 ppm Fe, 200 ppm Ni), which play a key role in the photooxidation of P3HT (Dupuis et al. 2012). By performing the fluorescence emission spectroscopy on films with different impurity concentrations, it is demonstrated that these impurities, even at low concentrations, have a catalytic effect on the decomposition of hydroperoxides into photoproducts that quench the fluorescence of P3HT by acting as electron traps and reduce the carrier transfer between P3HT and electron acceptor material (such as PCBM), and hence the performance of P3HT-based devices.

Some metallic impurities introduced to the material by this process may behave in a similar way as those that exist in conventional semiconductors and act as charge carrier trapping, even at very low concentrations. Among the common contaminates from reagents, palladium is one of the most frequently found and analyzed.

Trace impurities of palladium (Pd) in PTB7/PC71BM-based solar cells are thought to create charge carrier trapping centers, but are difficult to identify because of their very low concentration (< 1%) (Nikiforov et al. 2013). To probe the trapping transient absorption spectroscopy effects of Pd impurities, *transient absorption spectroscopy* (TAS) has been performed on PTB7/PC71BM films. This technique consists of using ultrashort laser pulses as a pump light to excite the sample and a probe light to measure the transmittance. The probe light is delayed by a controlled time relative to the pump light, and the transient absorption at various delay times is recorded. By comparing the decay times of the transient absorption spectra for pristine PTB7/PC71BM and 5% added catalyst $Pd(PPhh_3)$), it is shown that Pd impurities are responsible for the loss of charge carriers to recombination through the trapping process. By correlating the average distance between Pd atoms and the measured exciton diffusion distance, it is suggested that traces of Pd impurities act as trapping centers, even in pristine PTB7/PC71BM materials. The concentration of Pd in PTB7 is estimated at 0.16 $mg.cm^{-3}$.

On a larger scale, mixing or blending an organic semiconductor with one or more organic or inorganic materials is also a process that can produce defect centers in a material. The guest/host concentration ratio of the blend strongly affects the performance and stability of the material and the defect state distribution.

The effects of the blending of a conjugated polymer MDMO-PPV with different concentrations of PCBM ranging from 5 to 90% weight fraction have been studied by PDS (Goris et al. 2005). The absorption spectra of the MDMO-PPV:PCBM blends show an additional broad band ranging from 1.10 to 1.68 eV, compared with the individual spectrum of the two materials. This extra band is attributed to a *charge transfer complex* (CTC) formed by the interaction of the pure materials, which absorbs at low energies. The CTC is an electron donor–electron acceptor complex, which can be excited by an electron transfer from the donor to the acceptor. In the MDMO-PPV:PCBM blends, it is observed even at the lowest concentration of PCBM (5%) and interpreted in terms of defects and disorder in the conjugated polymer.

5.2. Defects in organic devices

Organic devices such as OLEDs, OPVs and OFETs are composed of several layers because of the need to be connected to the instruments or set-up to be used.

These devices consist of several layers, from the simplest structure of the electrode–semiconductor–electrode to a more complex structure of the electrode–hole transport layer–semiconductor–electron transport layer–electrode. To determine the defects in these devices, the whole structure should be considered since it is practically impossible to isolate the semiconductor layer in order to perform the measurements. As several layers are involved in the structure of a device, each of them may contribute to the measured defects, and their contribution should be taken into account. On the other hand, the interface between two adjacent layers of the device is prone to forming defects, especially when the morphology of the layer surface presents structural defects or impurities itself. Therefore, care should be taken in interpreting the defect measurements in organic devices, which usually need several experimental stages to obtain correct and reliable results, especially when using electrical characterization techniques such as TSC or DLTS.

Hereafter, we will consider the main factors that impact the introduction of defects to organic devices, which include defects from the semiconductors, from the interface layers and introduced by interlayer diffusion species. These defects are detected and characterized by different techniques previously described. It should be noted that electrical methods are generally used for characterizing defects in devices, since optical methods are rarely possible and applicable to multilayer structures.

5.2.1. *Defects from the semiconductor*

In an organic device, the semiconductor is the main active layer, and the measured defects are related to the semiconductor material. However, it should be stressed that when the defect measurements are performed by electrical techniques on devices, they can include defects from all the layers, that is, the electrodes, the transport layers and the semiconductors. For a basic diode structure, the interface layers between the electrodes and the semiconductor may introduce additional defects, and for a multilayer diode, additional traps can be introduced by any layer of the structure.

As an illustration, the characterization of traps in poly(9,9-dihexylfluoreneco-N,N-di(9,9-dihexyl-2-fluorenyl)-N-phenylamine) (PF-N-Ph)-based diodes has been investigated using the Q-DLTS technique (Nguyen and Renaud 2009). To identify the origin of the measured defects, three device structures were used: *ITO/(PF-N-Ph)/Ca/Al, ITO/PEDOT: PSS/(PF-N-Ph)Ca/Al*, and *ITO/PEDOT: PSS/(PF-N-Ph)/Al*. In diodes without the PEDOT:PSS layer, three trap levels are determined with activation energies of 0.13, 0.23 and 0.56 eV. Upon incorporation of the PEDOT:PSS layer, two additional traps with activation energies of 0.33 and 0.48 eV are identified, suggesting that they are introduced to the device by the transport layer

(Figure 5.1). Furthermore, when replacing the Ca/Al cathode layer by Al, no notable change in the trap parameters is observed. These results clearly indicate that the measured traps in the device studied have two different origins: traps due to the bulk of (PF-N-Ph) and traps due to the PEDOT:PSS layer.

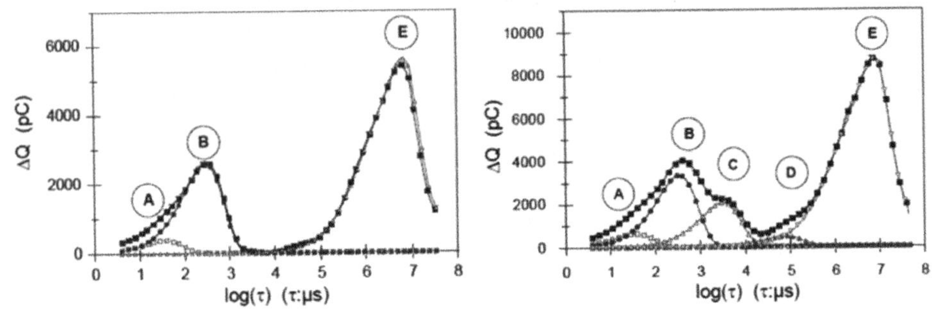

Figure 5.1. *Resolved Q-DLTS spectra of ITO/(PF-N-Ph)/Ca/Al and ITO/PEDOT:PSS/(PF-N-Ph)/Ca/Al diodes recorded at $T = 300\,K$, with a charging time of $t_c = 1\,s$ and a charging voltage of $\Delta V = 6V$ (Nguyen and Renaud 2009). For a color version of this figure, see www.iste.co.uk/nguyen/defects.zip*

5.2.2. Defects from the surface and the interface

The surface of an organic layer can have a significant density of defects, which is comparable to that of the bulk material. Moreover, depending on the preparation technique, the obtained layer can contain impurities from contaminants, which increases the defect density on the surface. In doped, blended or hybrid materials, the mixture of several organic and/or inorganic component leads to complex surface morphology, with the formation of domains limited by grain boundaries, which are a source of surface defects. At the contact of two adjacent grain boundaries, the change in the material structure can generate additional localized states acting as trapping centers and influencing the carrier recombination process.

In a device structure, multiple interfaces are formed by stacking different layers together. The property of each interface depends on the defect states and the morphology of the surface of each layer and influences the charge transfer across the interfaces. A well-known interface process in organic devices is energy level alignment at organic/organic or organic/inorganic interfaces, which is at the origin of the interface dipole formation (Crispin et al. 2002). It has been established that this process depends not only on the electronic properties of the two materials, but also on other factors such as surface defects, molecular orientation or environmental conditions. In the case of organic/organic heterojunctions, the contact between a small molecular layer and another organic material can influence the orientation of the molecules of the surface, locally interrupting the molecular order and affecting the charge transfer through the interface (Sehati et al. 2013). In devices such as OLEDs or OPVs, charge carrier injection or extraction is processed across the interface between the organic layers and the metal electrodes, and the charge recombination generally occurs at the vicinity of the interfaces. The performance of these devices clearly depends on the defect states at the interfacial region. Specifically, defects present at the interface between the oxide and the organic semiconductor strongly influence the electrical properties of OFETs, namely the transistor transfer characteristics in the subthreshold region (Scheinert et al. 2002).

The metal/organic (polymer) interface has been intensively studied because it strongly affects the charge injection into devices. For an efficient charge injection, the use of a high work function conductor is required to inject holes from the anode, and a low work function metal is needed to inject electrons from the cathode. The nature of the metal/organic interface has been investigated by several groups using XPS and UPS, and evidence of the interface states has been proved. These states can be introduced, namely when a metal is evaporated upon the organic layer, by the formation of the chemical bonds and defects between the organic semiconductor and the metal (Watkins et al. 2006) or of an organometallic complex (Nguyen et al. 2001). The interface gap states can pin the Fermi level, making the contact insensitive to the work function of the metal, and efficient charge injection can be obtained even when using high work function metal for the cathode, for instance. The conditions of the metal/organic interface formation are, however, still unclear and from experiments performed on the interface analyses and the device performance measurements, several hypotheses have been proposed including interface doping (Bharathan and Yang 1998), interface processing (Hwang et al. 2009), etc. By using the Q-DLTS technique, the interface trap parameters in P3HT:PCBM-based solar cells with metal/organic blend interfaces have been determined. The devices used are $ITO/(PEDOT{:}PSS)/P3HT{:}PCBM)/metal$ (Al and Ca/Al) and are identical except for the metal electrodes (Nguyen et al. 2014). At room temperature, the current–voltage characteristic of devices with low work function (Ca/Al) shows a higher current density, as compared to that of devices with high work function (Al), as expected from the energy diagram of the devices. However, the trap analysis of both devices indicates that if both devices contain defects from the P3HT polymer, an additional deep trap level of high density is present in the device with the Ca/Al electrode. This trap level is assigned to defects inherent to fullerene. In addition, the Q-DLTS spectrum of the $ITO/(PEDOT{:}PSS)/P3HT{:}PCBM)/Al$ device is comparable to that obtained in a P3HT-only device, and suggests that the deposition of Al on the blend would reduce or suppress defects generated by fullerene. It is suggested that a chemical reaction between Al and oxygen atoms of PCBM occurs and neutralizes the fullerene molecules at the interface (showing no fullerene defects), leading to a phase separation and a P3HT-rich interfacial region.

To investigate surface and interface defects of organic layers, several optical methods can be used. To determine the film structures, electron microscopy techniques such as *transmission electron microscopy* (TEM) and *scanning tunneling microscopy* (STM) allow molecular orientation and defects such as vacancies, interstitials, grain boundaries and impurities to be examined. For local investigation of the energy level alignment at the organic interfaces, *atomic force microscopy* (AFM) techniques such as *Kelvin probe force microscopy* (KPFM) (Qi 2011) can be employed to detect the formation of localized energy states. The principle of the KPFM technique is similar to the Kelvin method used to measure the work function of a metal surface. The KPFM signal is measured between the tip and the sample surface by applying a bias to nullify the *contact potential difference* (CPD). The method allows for mapping the surface potential of the organic layer and studying the formation of surface defects of the film, when it is exposed to different media under different exposure conditions (in the dark, under illumination, in air, under vacuum). Other analysis methods for studying the formation of the surface and the interface layer between the organic semiconductor layer and other materials such as X-ray photoemission spectroscopy (XPS) and ultraviolet photoemission spectroscopy (UPS) are also employed to determine the chemical composition and chemical reactions occurring on the surface and in the interfaces of the devices.

As an illustration of investigations of surface and interface defects of organic materials, organometal halide perovskites have been intensively studied for their application in organic solar cells. Here, we present a short summary of the results obtained from different studies using the techniques previously mentioned. On the surface morphology of perovskite thin films, polycrystalline grains are formed with variable sizes (~50–500 nm) depending on the deposition conditions and methods. Adjacent grains may have different crystal orientations, structures and compositions. The grain boundaries are found to be at the origin of the hysteresis observed in solar cells (Shao et al. 2016) by the accumulation of mobile ions in the vicinity of the grains. Furthermore, the charge accumulation forming a potential barrier at the grain boundaries affects the charge transport across the grains. The energetic distribution of traps on the surface of the $CH_3NH_3PbI_3$ perovskite is estimated by UPS measurements and covers a range from 100 to 400 meV below the band gap (Wu et al. 2015). These traps are assigned to the self-trapping of band-edge excitons.

On the interface between the functional layers, the position of energy levels in perovskite layers and their interfaces has been determined by UPS, for occupied states, and IPES (inverse photoemission spectroscopy) for unoccupied states (Schulz et al. 2014). The heterojunction band diagrams of perovskite in contact with some transport layers have been established, providing information and explanations of the interface characteristics, such as band bending and interface states.

Concerning the OFETs, it is often observed that the carrier mobility depends on the gate voltage, and this dependence is explained by the presence of trapping centers at the interface dielectric/organic semiconductor. Electrical characterization of devices using different interface structures by changing the dielectric materials and keeping the same organic semiconductor makes it possible to verify this assumption and determine the interface defect density. For instance, interface defects in OFETs using single crystals of α-quaterthiophene (4T), α-sexithiophene (6T), pentacene and tetracene, as well as thin films of pentacene have been determined by this technique (Schön and Batlogg 2001). Two different transistor structures have been realized using the same organic semiconductor (6T) with different gate dielectrics (Al_2O_3, Kapton). After verification that the bulk trap densities of both devices are similar (from space charge-limited current measurements), the plots of the subthreshold characteristics of the devices clearly show the influence of the interface traps at the semiconductor/dielectric interface. To determine the interface trap density N_{it}, which is related to the interface capacitance $C_{it} = |e|N_{it}$, the *subthreshold swing* S is written for devices with significant interface trap density (Sze 1981)

$$S = \frac{kT}{|e|}(ln10)\left(1 + \frac{C_D + C_{it}}{C_{ox}}\right) \qquad [5.1]$$

where C_D is the capacitance of the accumulation layer and C_{ox} is the capacitance of the dielectric layer. By measuring the subthreshold swing in both devices, with ($C_{it} > 0$) and without ($C_{it} = 0$) interface trap states, the interface trap state density in the Kapton device is estimated to be $N_{it} \sim 7 \times 10^{11} cm^{-2}$.

Because of the primary role of surface and interface defects in the charge transport, and hence in the performance of organic devices, several strategies have been proposed, which are aimed at reducing the defect density localized on the surface and in the interface layers. Among these strategies, *passivation* is one of the most used in organic devices to neutralize the active defects present on the surface layer or in the interfaces. This technique has been used to minimize the surface recombination rate in silicon-based solar cells by deducing the surface state density (Aberle 2000). Several methods have been applied with success, including growth and/or deposition of a dielectric film (thermal SiO_2 film), chemical passivation to saturate Si dangling bonds at the surface, high-low junction, etc. In organic semiconductors, namely in organic solar cells, the passivation strategies are similar to those used for silicon-based devices (Fu et al. 2020). The main passivation methods will consist of (i) filling (vacancies) or fixing (interstitials) the defects by chemical interactions, (ii) using a thin wide band gap layer to encapsulate the organic layer surface and (iii) improving the contact between layers through the energy level alignment. These methods can also apply to the volume of the organic semiconductor and the different transport layers used in the device structure.

Taking the case of perovskite-based solar cells as an example, the possible passivation processes of these devices can include several engineering aspects: metal electrode/HTL interface, perovskite/ETL interface, grains in the perovskite volume, composition of the perovskite (Akin et al. 2020). These approaches aim to passivate the defects on the surface, at the grain boundaries, in the interface layers and in the volume of the perovskite materials by reducing the recombination centers and suppressing the ion migration. This can be achieved by diverse operations such as using (i) appropriate materials for transport layers, (ii) spacer interlayers for preventing ion diffusion, (iii) additive molecules for neutralizing trap states, (iv) appropriate substrates for reducing the grain boundary defects, (v) mixed-cations or mixed-halide perovskites, etc. The passivation of devices not only improves the transport properties by reducing the trap density and effects, thus enhancing the device performance, but also improves their thermal and environmental stability.

5.2.3. *Defects from diffused impurities*

In organic devices, we distinguish between two kinds of diffused impurities. The impurities that originated from the active organic semiconductor (intrinsic diffused impurities) and those that originated from other layers or media (extrinsic diffused impurities). Both of them influence the transport process since they act as trapping centers, just like the aforementioned chemical impurities.

Intrinsic diffusion usually occurs in composite or blended materials whose structure is loose and allows the weakly bound atoms to diffuse slowly through the material bulk. In bulk heterojunction (BHJ) solar cells using the P3HT:PCBM blend as an absorber, the interdiffusion of P3HT and PCBM has been observed in different thin film configurations (blend films and bilayer) upon thermal annealing, leading to a phase separation between the polymer and the fullerene, and thus the formation of nanosized domains of low concentration of PCBM (Lee et al. 2011). The diffusion of PCBM into P3HT is made possible through the grain boundaries that are formed in the polymer. Similarly, ion migration is intrinsic to MAPI perovskite materials because of the presence of constituting ions such as I^-, Pb^{2+} and MA^+, which can move from a lattice site to another by different pathways created in the material structure (deQuillettes et al. 2016). These pathways can be formed through lattice distortions by accumulated charges or impurities, or by open space at grain boundaries (Yuan and Huang 2016). The ion migration occurs under the influence of an electric field or under a light illumination, producing the diffusion of charged point defects, which affects the internal electric field distribution, and are thought to be at the origin of many observed phenomena in perovskite devices such as: current–voltage hysteresis (Tress et al. 2015), photoinduced phase separation, conversion between lead iodide PbI_2 and $MAPbI_3$ (Yuan et al. 2016), etc. To improve the stability and efficiency of perovskite-based devices, the ion migration should be minimized by restricting the available migration pathways, especially by enhancing the grain size to reduce the ion diffusion.

Extrinsic impurities diffuse into the organic semiconductor layer of devices through its contact with the electrodes. The diffused species can be oxygen or water when the devices are exposed to air or a humid atmosphere, but metal atoms from the electrodes (anode or cathode) can also diffuse into the organic layer.

Water vapor and oxygen enter the organic semiconductor through defects in the barrier materials (pinholes in the electrode, grain boundaries formed on the layer surface) (Chatham 1996) and act as an electron trap. Consequently, it favors exciton quenching and generally lowers the electrical conductivity of the material. Once it has reached the surface of the organic material, oxygen/water will oxidize the sublayer and possibly diffuse through the whole film and the layers, causing the degradation of the material and the device.

Most of the organic devices have a transparent electrode (usually ITO) for exchanging photons with the external media and metal atoms enter the semiconductor by diffusion from the ITO anode, which releases In and/or Sn ions upon dissolution of the oxide surface by etching (Nguyen et al. 2001; Lee et al. 1999). As an illustration, Figure 5.2 shows the XPS profile evolution related to aluminum, carbon, oxygen and nitrogen in the Alq3/Al interface (top contact) obtained in a diode of structure $ITO/Alq_3/Al$. After failure of the diode, it can be seen that indium from the ITO electrode has diffused through the organic layer (100 nm thick) and has reached the top electrode. The chemical degradation of the ITO electrode is greatly facilitated by the use of PEDOT:PSS as a hole transporting layer because of its hygroscopic nature. Moreover, the PEDOT:PSS itself with excess PSS concentration is possibly degraded in devices, and PPS can diffuse to adjacent layers to form compounds modifying the initial structure of these layers (Norrman et al. 2006).

From the cathode side, metal electrodes can penetrate the organic layer upon evaporation or undergo electrochemical reactions after deposition, liberating metal ions that diffuse through the metal/organic interface and enter the organic layer. In the first case, when metal atoms are deposited on the organic film, the released energy is sufficient to break the weak van der Waals bonds of the organic surface and allow them to diffuse into the organic sublayers. The diffusion depth depends on the nature of the metal and the structure of the organic film, and in some cases does not exceed a few nanometers (Lee et al. 1999). In the second case, when a metal alloy such as Mg:Ag is deposited on an organic surface, a chemical reaction usually occurs between the organic film and the alloy, decomposing the latter and liberates ions of one of the metal components that will diffuse into the organic film. The chemical reaction may also occur when a metal/alkaline metal compound such as

Al/LiF is deposited on an organic film. It takes place between the metal and the alkaline metal compound and liberates alkaline ions, allowing them to diffuse into the organic film (Heil et al. 2001).

Diffusion of metal atoms or ions in organic devices has been observed when depositing metal layers such as Al (Turak et al. 2002), Mg (Rajagopal and Kahn 1998), Ag (Song et al. 2001), Au (Seki et al. 2001) and Li (Parthasarathy et al. 2001) on the organic films (small molecules or conjugated polymer). One possible advantage is that the metal diffusion is an enhancement of the injection of charge carriers from the related electrode to the organic semiconductor, which act as dopants (Heil et al. 2001). However, in most cases, the injected metal atoms or ions tend to react with weak bonds of the organic structure, which in turn alter the device structure layers and affect its operational stability.

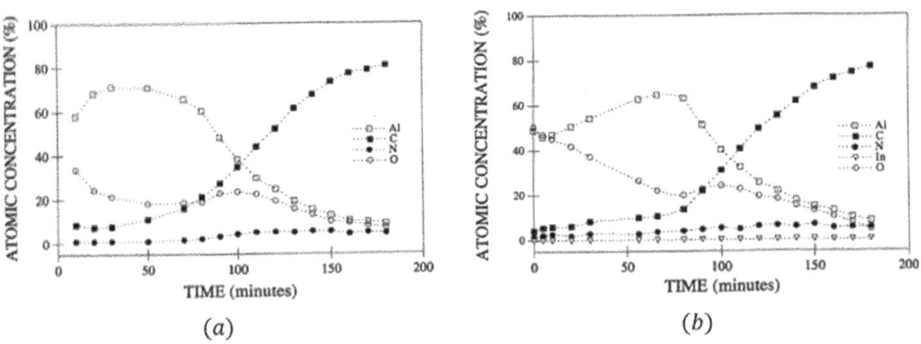

***Figure 5.2.** XPS profiles of aluminum, carbon, nitrogen, oxygen and indium in the Alq3/Al interface (top contact of an ITO/Alq3/Al diode): a) before, b) after degradation, showing indium diffusion from the ITO substrate (Nguyen et al. 2001, p. 75)*

Diffusion of extrinsic ions can occur through the contact formed between a charge transport layer with the organic semiconductor, especially when the transport layer is intentionally doped to enhance the carrier injection. In perovskite-based devices, the extrinsic diffusion of Li^+ ions has been observed when using spiro-OMeTAD as an HTL, doped with Li (Li et al. 2017). The ions move through the perovskite layer and then reach and accumulate at the opposite electrode, which modifies the charge at the interface ETL/perovskite region.

In order to reduce the diffusion of impurities into the organic materials, several methods have been proposed and tested. To limit the diffusion of oxygen or water from the outer media, a thin film encapsulation of the device is usually efficient to slow down the diffusion rate. The encapsulation material should be composed of an alternating stack of inorganic layer (acting as a diffusion barrier) and organic layer (used as a flexible matrix), and to obtain a protection of high quality, the required *oxygen transmission rate* (OTR) and *water vapor transmission rate* (WVTR) should be as low as $10^{-5}cm^3m^{-2}day^{-1}$ for OTR and $10^{-6}g.m^{-2}day^{-1}$ for WVTR (Ahmad 2014). The main difficulty in realizing the inorganic/organic stacked thin films is the potential damage caused by the deposition of the inorganic layer, which requires high-energy conditions. Low-temperature methods such as *atomic layer deposition* (ALD) can be used to prevent the organic layer damage during the deposition.

To limit the diffusion of metal atoms or ions from the ITO electrode, the use of thin passivation layers such as copper phthalocyanine (CuPc) (Van Slyke et al. 1996) or copper oxide (CuO_x) (Hu et al. 2002) has been proven to efficiently stabilize the diffusion. Other treatments of the ITO surface have also been successfully applied to minimize the effects of metal diffusion like UV–ozone treatment (Ishii et al. 2000) or insertion of a self-assembled layer between the transport layer and the electrode (Wong et al. 2002). To prevent the diffusion from metal electrodes other than ITO, a thin buffer layer inserted between the organic active layer and the electrode (anode or cathode or both) has proven to be efficient in reducing the metal diffusion and enhancing the device stability. The materials used for the buffer layer can be metal oxides (MoO_3) (Deng et al. 2021) or organic materials Alq3, Bphen (Angel et al. 2017).

6

Defects, Performance and Reliability of Organic Devices

The ultimate objective of investigating any electronic device (organic or inorganic) is its commercialization. For this aim, several essential and special criteria are required for the marketable products. Indeed, customers would purchase a good when they find a purpose and a need for that good. On the other hand, they will expect that the quality of the good and its service life would be reasonable in relation to the price paid for it. As an illustration of this statement, let us consider the commercialization of an organic solar cell. When customers consider purchasing organic solar cells for environmental or economic reasons, they would choose the cells with the highest efficiency for obtaining maximum energy from the sun. They would also expect to use the cells during a long period (of many years or decades) before replacing them. Finally, they would buy these cells if their price was reasonable and affordable. The potential cell manufacturers have to take these expectations into account when they commercialize organic solar cells, knowing that silicon-based solar cells are already available on the market. This does mean that the silicon cells address all the aforementioned criteria, and to be able to compete with them, organic solar cells should present advantages over their counterparts, at least in some aspects.

In the scheme proposed by Brabec (2004), the requirements for commercialization of organic solar cells can be represented by a critical triangle with three criteria: lifetime, efficiency and cost. Each of them occupies a corner of the triangle. This representation can evidently apply to other organic electronic devices such as OLEDs and OFETs.

As for the reliability of these electronic devices, we understand that they should be able to perform their required function for a defined period of time under given conditions. For each type of device, in the operational characteristics being specified (technical performance, environmental conditions), reliability mainly concerns the functioning of the devices over time, that is, the evolution of their initial technical performance to the limit of an acceptable level. Therefore, the reliability of a device is intimately linked to its working time and the performance quality is usually specified at the initial time ($t = 0$).

In this section, we will examine the influence of defects on the efficiency (or performance) and the lifetime (or the operational stability) of organic devices, to emphasize the role of defect investigations in the future evolution of electronic equipment commercialization.

6.1. Impact of defects on the performance of organic devices

It appears evident that defects influence the quality of semiconductors and consequently that of the devices using these semiconductors as active layers. However, identification of the origin of defects in devices is not simple as there may be several factors to be considered, depending on the material synthesis, the device fabrication, the measurement technique used and the experimental conditions. Only when the origin of the defect creation is clearly established, can the relation defect performance of a device be investigated. Therefore, various strategies have been proposed to study the influence of defects on the quality of devices in order to improve their reliability, which is essential for the production and commercialization. A convenient way of relating the device performance to the semiconductor defects consists of improving the quality of the active material and then observing the enhancement of the device performance. For instance, the purification of conjugated polymers is commonly used for obtaining high-quality material with reduced defect density for the active layer and for improving the device performance (Craciun et al. 2010). In contrast, it is also possible to alter the performance of a device by intentionally introducing impurities that increase the overall defect density of the semiconductor. For instance, exposition of a device in air would allow a diffusion of oxygen and water to the active layer, which will alter its structure and create defects, leading to a decrease in the device performance (Khelifi et al. 2011).

6.1.1. *Defects and efficiency of OLEDs*

A basic OLED is composed of a semiconducting layer sandwiched between two electrodes. The anode is a transparent conducting layer (usually ITO), which allows extraction of the emitted photons in the semiconductor and the cathode is a metal layer of low work function, which facilitates electron injection into the active layer. In more elaborated diodes, charge transport layers are inserted between the emitting layer and the anode (hole transport layer HTL) or the cathode (electron transport layer ETL) with increased injection charge density. Under an applied voltage, electrons and holes are injected from the cathode and anode respectively. The charge carriers will move to the emitting layer where they form excitons, which decay to the ground state, and photons are emitted to the external medium.

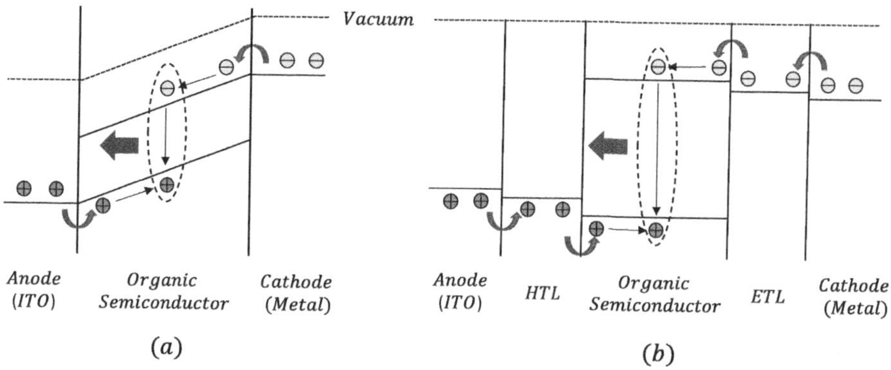

Figure 6.1. *a) Physical processes of light emission in basic OLED structure: charge injection, charge transport and electron–hole recombination. b) Physical processes of light emission in multilayer OLED structure comprising transport layers HTL and ETL. For a color version of this figure, see www.iste.co.uk/nguyen/defects.zip*

The performance of an OLED can be measured by the external quantum efficiency (EQE), which is defined as the number of photons generated to the external medium per number of charge carriers injected to the device. The EQE can be expressed by

$$\eta_{EQE} = \eta_{op} \times \eta_{IQE} \quad [6.1]$$

where η_{op} is the optical out-coupling efficiency and η_{IQE} is the internal quantum efficiency. The η_{op} efficiency depends on the optical properties of the materials used in the device, while η_{IQE} efficiency depends on the injected charge carriers and the exciton–photon conversion efficiency.

η_{IQE} represents the ratio of the number of emitted photons to the number of injected carriers into the diode, that is, the efficiency of converting injected charges into photons. It can be written as

$$\eta_{IQE} = \gamma\, \eta_{e/p} \Phi_{PL} \quad [6.2]$$

where γ is the charge carrier (electron and hole) balance factor, $\eta_{e/p}$ is the exciton–photon conversion efficiency and Φ_{PL} is the photoluminescence quantum yield. Other forms of efficiency OLEDs are the current efficiency and the luminous efficacy, which take into account the sensitivity of the human eye and are not used here to relate the effects of defects.

Generally speaking, in the presence of defects in a semiconductor, the charge transport in the material is affected by the effects on the carrier mobility and free carrier density. On the one hand, the movement of free carriers is influenced by the Coulomb interaction between the carriers and the defects, resulting in a deflection of their paths and a change in their mobility. On the other hand, if deep trap states are present in the band gap, the carriers trapped by these traps can no longer contribute to the conduction process in the semiconductor, and its overall conductivity will be reduced. Finally, defects at the device interface can affect the charge injection, which controls the density of free charge carriers in the semiconductor.

To detect the trapping process in OLEDs and relate it to the performance of devices, several experimental techniques can be used: electrical and optical measurements such as TSC, DLTS and PL for the defect characterization and current–voltage I(V), luminance–voltage (L–V) for the performance characterization. The correlation between defects and performance can be achieved by a comparison of the defect parameters and the performance characteristics of devices (Hepp et al. 2003b).

The presence of defects in an OLED structure influences its overall efficiency by affecting the η_{IQE} efficiency: (i) by creating or enhancing the charge imbalance between hole and electron injection by defects at the electrode surface and defect diffusion from the electrodes. The charge imbalance can also be caused by defects at the interfaces of the transport layers incorporated into the device, (ii) by creating non-radiative decay paths, which reduce the PL quantum yield. Impurity defects such as halogen or oxygen atoms introduced to the organic emitting material by synthesis or diffusion act as luminescence quenchers, non-radiative recombination centers or deep charge traps (Giebink et al. 2008), and reduce the light output of the OLEDs.

6.1.2. *Defects and efficiency of OPVs*

A basic organic solar cell is composed of a transparent conducting electrode (usually ITO), an organic active layer (*absorber*) and a metal electrode. When exposed to light, the cell converts the light energy into electrical energy through a conversion process with four stages: (i) absorption of a photon that creates an exciton, (ii) diffusion of the created photon to a recombination site, (iii) dissociation of the exciton into a hole and an electron, (iv) transport of charges to the electrodes and creation of a voltage at the terminals of the device.

In process (i), the created excitons can be geminate (or an initial pair of charged carriers) or non-geminate (a pair of charged carriers formed after diffusion) (Mort 1981). The photogeneration efficiency of geminate recombination is field and temperature dependent. Non-geminate recombination may be *trap-assisted recombination* (Shockley–Read–Hall or SHR) or *bimolecular recombination* (Langevin), and is diffusion dependent (the Auger recombination being not dominant). The recombination processes affect the loss of photogenerated carriers and consequently, the reduction of the solar cell efficiency.

The PCE of organic solar cells is defined as the ratio of the generated electric power P_m to the incident light power P_{in} on the solar cell, and can be expressed by

$$\eta = \frac{P_m}{P_{in}} = \frac{J_m V_m}{P_{in}} = \frac{J_{SC} V_{OC} FF}{P_{in}} \qquad [6.3]$$

where J_m and V_m are the maximum current density and voltage respectively, J_{SC} is the short-circuit current density, V_{OC} is the open-circuit voltage and FF is the fill factor.

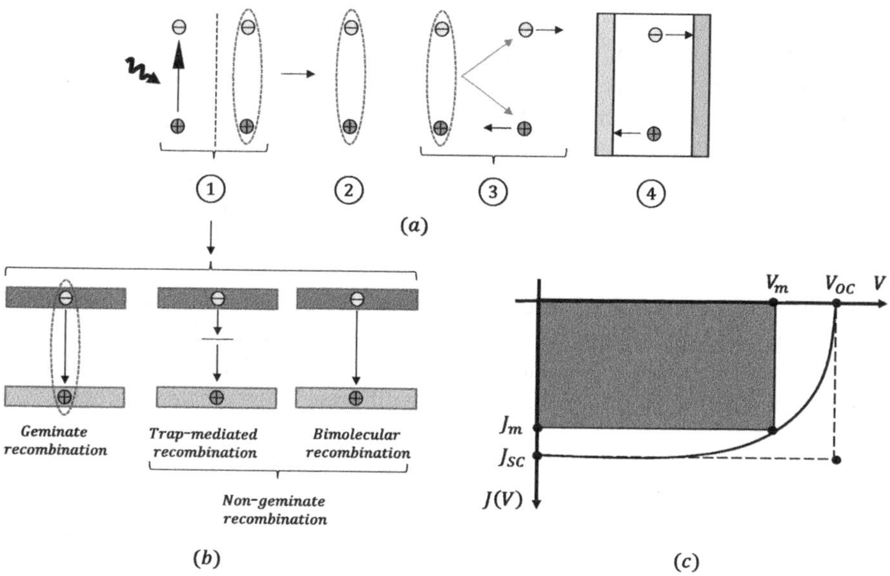

Figure 6.2. *(a) Working principle of organic solar cells. (b) Geminate and non-geminate recombination. (c) Schematic representation of current–voltage (J–V) characteristics of solar cells under illumination. For a color version of this figure, see www.iste.co.uk/nguyen/defects.zip*

In equation [6.3], J_{SC} is determined by the external quantum efficiency η_{EQE}, which represents the fraction of photogenerated charges collected at the electrodes of the solar cell. According to the conversion process described previously, it can be written as

$$\eta_{EQE} = \eta_A \times \eta_D \times \eta_C \quad [6.4]$$

where η_A is the light absorption efficiency, η_D is the exciton efficiency and η_C is the carrier collection efficiency. To obtain a high short-circuit current density, the absorber should have a high photon absorption, a good carrier transport and charge collection properties.

On the other hand, V_{OC} depends mainly on the difference in energy between the HOMO level of the donor and the LUMO level of the acceptor (Brabec et al. 2011)

$$E_{gap} = E^{D}_{HOMO} - E^{A}_{LUMO} \quad [6.5]$$

A high open-circuit voltage requires a high value of E_{gap}, which is determined by the choice of the donor and acceptor materials. Any modification of the HOMO and LUMO energy levels will affect the V_{OC} voltage.

Finally, the fill factor FF is a function of V_m and J_m and represents the intrinsic quality and properties of the absorber for its ability to minimize the recombination losses. It is shown that this factor depends on two intrinsic characteristics of the solar cells, which are the shunt resistance R_{sh} and the series resistance R_s of the equivalent circuit. The former is related to the leakage current in the device, which is linked to the imperfections of the materials composing the cell. The latter is related to the contact resistance, which is linked to the interfaces formed between different layers.

Obviously, the presence of defects in organic devices affects their performance because they influence the different factors of the efficiency through the carrier transport processes. Based on equation [6.3], the effects of defects on the solar cell efficiency have been analyzed in devices using different materials as absorbers.

6.1.2.1. *Defects and efficiency of bulk heterojunction (BHJ) solar cells*

In bulk heterojunction (BHJ) solar cells, the semiconductor blend is composed of donor–acceptor (D–A) domains, which allow for the dissociation of excitons into free electrons and free holes. These excitons are created in the donor or acceptor phase in the vicinity of the D/A interface by charge transfer from the donor to the acceptor. The free charge carriers then move to the electrodes and are collected to create a voltage at the device terminals. However, if trap states are present, the free charge carriers can be partly trapped and the density of collected charges will be reduced, leading to a decrease in the efficiency of the devices. On the other hand, recombination of the free carriers can also occur during their transport inside the semiconductor by different processes, diminishing the charge dissociation efficiency. Several approaches have been proposed to correlate the performance of BHJ solar cells and the defects formed in the devices.

The open-circuit voltage V_{OC} of BHJ solar cells depends on the balance between charge generation and recombination, which involves several parameters (Blaksley and Neher 2011): (i) illumination intensity, (ii) the donor–acceptor gap, that is, the energy difference between the HOMO level of the donor and the LUMO level of the acceptor, (iii) charge carrier recombination rate, (iv) contact work function and (v) the amount of energetic disorder. In the presence of defects, recombination between free holes in the donor (electrons in the acceptor) with trapped electrons in the acceptor (holes in the donor) occurs by trap-mediated recombination. The effects of traps on the performance of the solar cells can be verified by introducing carrier traps in the absorber, then measuring the variation of the V_{OC} voltage (Mandoc et al. 2007a). The results obtained for the open-circuit voltage of MDMO-PPV and PCBM solar cells proved that it is strongly affected by recombination of free holes with trapped electrons (trap-assisted recombination). This mechanism added to the Langevin recombination leads to an increase in the charge carrier recombination rate and thus to a reduction of the open-circuit voltage. The effects of the trap-assisted recombination are usually observed by an enhancement of the dependence of the V_{OC} on the light intensity.

In these devices, the short-circuit current J_{SC} and the fill-factor FF of BHJ solar cells, which are also the parameters in the expression of the PCE, are found to be dependent on the incident light intensity and are explained by the space-charge formation in the device, due to the difference in mobility of electrons and holes (Koster et al. 2005).

When the distance between the exciton charges is small, the Coulomb binding energy is high and reduces the dissociation efficiency of excitons. This reduction is observed in the polymer blend of MDMO-PPV- and PCNEPV-based solar cells (Mandoc et al. 2007b). Consequently, both J_{SC} and FF will decrease at operating voltages of the solar cells. In these devices, the recombination that contributes to the recombination losses of the devices are trap-assisted and Langevin recombination. Trap-assisted recombination would occur only at low light intensity, while Langevin recombination would be dominant at higher incident light intensity.

In BHJ solar cells using blends of PTB7:PC71BM as an absorber, the effects of defects are investigated by varying the donor (PTB7)/acceptor (PC71BM) ratios and studying the changes occurring in the efficiency parameters V_{OC}, J_{SC} and FF under AM 1.5G illumination (Ho et al. 2017). A clear correlation between PCE, J_{SC} and FF has been obtained through their similar variation with increasing acceptor concentration. The transport properties of the devices were studied using the SCLC model with traps and analyzed using the Gaussian disorder model (GDM), which yields the carrier mobility by the following expression:

$$\mu = \mu_0 \exp\left\{-\left(\frac{2}{3}\frac{\sigma}{kT}\right)^2\right\}\exp\{\beta F^{1/2}\} \qquad [6.6]$$

where σ is the disorder parameter. It is demonstrated that the electron transport is the main factor that influences and correlates the device efficiency, and consequently, the short-circuit current density J_{SC} and the fill factor FF. In particular, it is also shown that the fullerene acts as a trap and hinders electron mobility. When the fullerene concentration increases over a limit (percolation), the carrier mobility increases and the efficiency PCE of the device, as well as J_{SC} and FF decrease, suggesting that trapping is the key factor of the device performance.

6.1.2.2. *Defects and efficiency of perovskite solar cells*

In perovskite solar cells, the most predominant recombination mechanisms are trap-mediated and band-to-band for low and medium photoexcitation density. Under solar illumination, the *external electroluminescence quantum efficiency* EQE_{EL} is directly related to the open-circuit voltage V_{OC} (Rau 2007), and the competition between the non-radiative trap-mediated and radiative band-to-band recombination will affect the V_{OC} voltage value, which can be expressed by (Stranks 2017)

$$V_{OC} = V_{OC}^{max} - V_{OC}^{non-rad} = V_{OC}^{max} - \frac{kT}{|e|} \times |ln(\eta_{LED})| \qquad [6.7]$$

where η_{LED} is the EQE of the device working as an LED.

From this equation, a high defect density and a high non-radiative combination rate will reduce the V_{OC} voltage of the solar cell since they contribute to the non-radiative voltage loss $V_{OC}^{non-rad}$.

Under steady-state illumination, electrons are excited from the valence band to the conduction band, splitting the quasi-Fermi levels E_{Fn} and E_{Fp}, which determines the open-circuit voltage $V_{OC} = E_{Fn} - E_{Fp}$. When shallow traps are present, the quasi-Fermi levels of perovskite can be pinned by energy-level alignments at the contact of the perovskite with adjacent layers (ETL and/or HTL). Consequently, the V_{OC} voltage can be reduced, especially when additional non-radiative recombination processes are introduced by defects.

Generally, as described previously, the perovskite devices have a high defect density, and charge carrier trapping leads to a low value of V_{OC} voltage by non-radiative decay from the defect sites. However, a study of kinetics in perovskite material and devices has shown that the recombination pathways of trapped charge carriers depend on the excitation intensity and sample temperature (Stranks et al. 2014). At low fluences or high temperature, electrons and holes tend to stay free and can be trapped by defect sites, resulting in non-radiative decay. In contrast, at high fluences or low temperature, they tend to be bound as excitons and increase the radiative processes.

As in BHJ structures, correlation between the carrier trapping process and the performance of the perovskite solar cells is studied by measuring the efficiency parameters (V_{OC}, J_{SC} and FF) of the devices in which the trap density of the absorber (or of other layers) is modified by diverse techniques. Often, the correlation defect/performance is also analyzed by the *hysteresis characteristic* obtained during the measurement of the current–voltage ($J - V$) curves. This characteristic represents the shift of the current measured in forward and reverse scanning directions, while sweeping an applied voltage between short-circuit ($V = 0\ V$) and open-circuit ($V = V_{OC}$) conditions to the device (Snaith et al. 2014). Several hypotheses have been proposed to explain the origin of hysteresis in perovskite: defects at the surface and interface of the perovskite film (Kim et al. 2013), modifications of energy-level alignment at the perovskite interfaces (Baena et al. 2015). It seems accepted that the hysteresis in perovskite solar cells originated from defect ion migration involving mobile ions, which accumulate at the interfaces and create dipole layers, enhancing the electric field at the contact (Tress 2017). As a consequence, a large measured hysteresis characteristic of devices reflects a high ionic defect, which can trap the carriers and lead to an increase of defect-mediated recombination, and therefore a low performance of the solar cells.

6.1.3. *Defects and performance of OFETs*

The working principle of a basic OFET is based on the creation of a channel in a portion of an organic semiconductor placed between two metal electrodes S (the source) and D (the drain) (Figure 6.3).

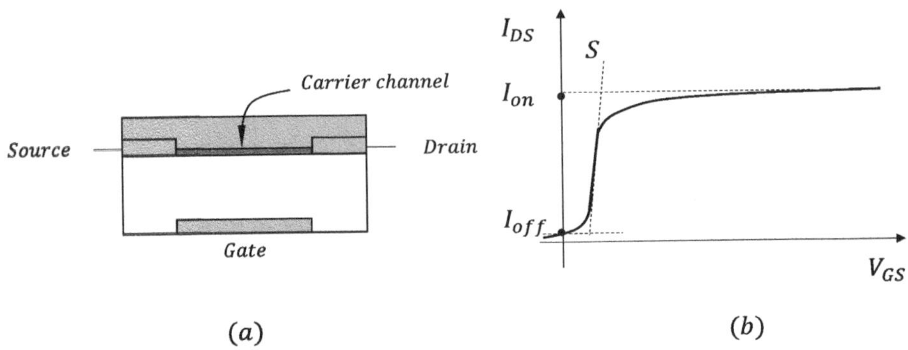

Figure 6.3. *a) Schematic OFET structure. b) Schematic transfer characteristic $I_D(V_{GS})$ with switching parameters: I_{on}, I_{off} and subthreshold swing S. For a color version of this figure, see www.iste.co.uk/nguyen/defects.zip*

The semiconductor is deposited on a dielectric film, which isolates it from a third electrode G (the gate). When a voltage V_{DS} is applied between the source and the drain, a *drain current* I_{DS} (or I_D) will flow in a channel between the electrodes. However, its intensity is weak because of the low conductivity of the organic semiconductor. By applying a voltage V_{GS} between the gate and the source, it is possible to modulate the conductance of the channel and change the current intensity I_D by the field effect. The drain current is a primary parameter for evaluating the performance of OFETs. It can be expressed by

$$I_D = \frac{1}{2}\frac{W}{L} C_i\, \mu (V_{GS} - V_T)^2 \quad [6.8]$$

where W and L are the channel width and length respectively, C_i is the gate capacitance per unit area, μ is the field-effect mobility and V_T is the threshold voltage. However, it is not used for comparison of device performance because its values depend on different parameters that vary from one sample to another. Instead, the field-effect mobility μ is generally employed to evaluate the quality of an OFET, since a high mobility allows for better potential applications of the device. Other characteristics are also used for comparison of OFET performance, such as the *threshold voltage* V_T, which defines the operating voltages of the device. The switching characteristics, which are essential for OFET applications used as a switch in logic control circuit, are also indicative of the device performance. For instance, the switching speed from the "off" to the "on" state is measured by the *subthreshold swing* $S = \frac{\delta V_{GS}}{\delta[log_{10}(I_D)]}$, and the switching ability is measured by the *on/off current ratio*, which allows us to distinguish between the "on" and the "off" states, that is, to indicate the ability of the transistor to shut down.

For OFETs, it is more appropriate to speak in terms of performance instead of efficiency to describe the rendering of the device. Indeed, depending on the function to accomplish, OFETs are used differently, and the operation result is not necessary evaluated and expressed by a measured efficiency. In most cases, the determination of the field-effect mobility is used to prove the high performance of the studied OFETs.

The presence of traps in the organic semiconductor obviously influences the performance of OFETs since it reduces the carrier mobility, as explained previously. For organic semiconductors, the mobility is generally low as compared to that in inorganic counterparts. High field-effect mobility is of the order of 20 $cm^2V^{-1}s^{-1}$, as measured in rubrene single crystals (Podzorov et al. 2004), which is far lower than electron mobility in crystalline silicon (~1,500 $cm^2V^{-1}s^{-1}$). As in OLEDs, a purification of the organic semiconductor improves the mobility (or the performance) of the device by reducing the impurity concentration, which acts as a trapping center. For example, in pentacene-based devices, the main impurities are pentacenequinone, which can be eliminated by vacuum sublimation under a temperature gradient (Jurchescu et al. 2004). The OFETs made with purified pentacene have a field mobility of 35 $cm^2V^{-1}s^{-1}$ (at room temperature), which is better than that measured in OFETs using unpurified pentacene (~1 $cm^2V^{-1}s^{-1}$).

Further investigations and measurements have suggested the predominant role of trapping at the interface dielectric/organic semiconductor on the performance of the transistor (Veres et al. 2003). It is observed that the measured carrier mobility does not only depend on the organic semiconductor, but also, and strongly, on the dielectric layer/organic interface. The mobility tends to decrease when the relative dielectric constant ε_r increases. The effects are explained by an enhancement of the carrier localization, leading to a broadening of the DOS, more or less large depending on the dielectric constant ε_r, due to the dipolar disorder at the organic/dielectric interface. For example, OFETs using single crystal rubrene as a semiconductor with different dielectrics as gate insulators have field-effect mobilities varying from 1.5 $cm^2V^{-1}s^{-1}$ (with Ta_2O_5, $\varepsilon_r = 25$) to 10 $cm^2V^{-1}s^{-1}$ (with parylene C, $\varepsilon_r = 3.15$) (Stassen et al. 2004).

6.2. Impact of defects on the stability of organic devices

In electronics, the term "failure" is often used to refer to degraded or malfunctioning devices and components. If these devices or components are able to operate for a given time period without failure in predefined performance conditions, they are considered as "reliable" and the word "reliability" is employed to express the probability of operation of such devices or components. The elapsed time during which the operation of the device is carried out in normal conditions is termed as the *time to failure* (or *lifetime*). During this time, the device is working correctly with a stable output. After this period, the performance of the device changes and generally decreases.

Failure analysis aims to identify the causes of device failure by using spectroscopic techniques to detect the changes in physical and chemical properties of the semiconductor and the device, in order to understand the physical mechanisms that lead to the observed failure. For semiconductor materials and devices, in particular in the organic field, the physical mechanisms of failure are often called *degradation mechanisms*. When a semiconductor or a device is degraded, an identifiable change in their operational characteristics will be clearly observed and when investigating the degradation causes, we can apply the procedures described previously for the device performance analysis. Indeed, a comparison of some specific physical properties of the semiconductor or devices before and after degradation may be helpful in understanding the causes of failure. It should be noted however that the time to failure for electronic devices is generally long and the analysis is often time-consuming. On the other hand, because of the various chemical compositions and structures, there are numerous possible causes of failure in organic semiconductors and devices, much more than in conventional devices. Moreover, among these degradation mechanisms, competing failure modes (or electrical symptoms) may exist and a fine analysis is required to identify the dominant failure mechanism.

6.2.1. *Overview of degradation mechanisms in organic semiconductors and devices*

The methods for investigation of the failure causes in inorganic devices have been intensively used to study optical and electron devices (Fukuda 1991). For organic devices, the same methods have been applied, and degradation analysis concepts from inorganic electronics are also widely used to interpret the physical failure mechanisms.

6.2.1.1. *Degradation modes*

Depending on the origin of the defect sources, the failure mechanisms can be divided into three categories: intrinsic, extrinsic and electrical stress degradations (Ohring and Kasprzak 2015).

Intrinsic degradations are due to effects of defects and failure sources in the materials that compose the device, including the active semiconductor and other materials that are incorporated into the device to provide a high operation performance. These defects are intrinsic characteristics of materials, but can also be introduced to the structure during the manufacturing processes. In other words, the original materials and devices are not free of defects, but their concentration is generally low and does not affect the device operation and performance.

Extrinsic degradations are due to defects and failure sources induced by external triggers, which interact with the material and device. There are several extrinsic degradation triggers such as water, oxygen, light and heat, whose effects are commonly observed in organic devices. This means that these effects can be mediated by minimizing the interactions of the medium with the materials, which is not always possible in practice. The main sources of stress that lead to the premature aging of the devices are light irradiation, temperature, environmental media and electrical bias (in normal operation conditions). Fast degradation occurs in devices that are simultaneously exposed to several stress sources during their functioning.

Electrical degradation is caused by *overstressing* of the device during operation. It also occurs in normal applied bias conditions for aged devices whose structure has been previously weakened by intrinsic or extrinsic degradations.

These modes have been widely considered and accepted for inorganic electronic devices and they will be used in this section to describe the degradation of organic ones. It should be noted that organic devices may have other failure modes than inorganic systems due to their chemical structure and composition. Other mode classifications may be found in the literature, although they do not fundamentally change the nature of the degradation process.

6.2.1.2. *Degradation process characteristics*

Depending on the device function, degradation process can manifest through several characteristic changes. Evaluation of the device degradation versus time consists of determining its lifetime, that is, studying the change in performance over time. There are two main ways to study a device lifetime: (i) *storage lifetime*, which is the measured lifetime of a device when it has not yet been used in operation after its fabrication, (ii) *operational lifetime*, which is the measured lifetime of a device that is in operation after its first start-up. For practical use of devices, most of the reliability studies focus on determining the device operational lifetime.

The degradations of OLEDs are characterized by a loss of luminance over time, occurring during operation of the device in a predefined physical environment (constant temperature, applied voltage and humidity). Depending on the application field, the lifetime of an OLED is determined from the luminance decay curve as a function of time. For display applications, the lifetime T_x corresponds to a loss of $x = 50\%$ of the initial luminance (usually of $100\, cd.m^{-2}$), whereas for lighting applications, higher values of x are chosen (T_{70} or T_{80}). The decay curve for OLEDs (and for OPVs) can be described by an exponential function of time. When several physical processes are involved in the degradation of a device, each process can be represented by a function, and the global decay will be expressed as a sum of exponential functions. For example, a degradation involving two different processes can be described by the following expression for the luminance:

$$L(t) = L_0 \times [Aexp(-k_1 t) + Bexp(-k_2 t)] \qquad [6.9]$$

where L_0 is the device initial luminance at $t = 0$, A and B are constants, and k_1 and k_2 are the degradation rate constants.

It is sometime convenient to use a single function called the *stretched exponential decay* to describe the whole degradation behavior of the device by the following expression (Féry et al. 2005):

$$L(t) = L_0 \times exp\left[-\left(\frac{t}{\tau}\right)^{\beta}\right] \qquad [6.10]$$

where τ is the characteristic time and β is the stretching exponent. However, equation [6.10] remains mathematical and there is not a global consensus on the interpretation of the physical process using these simple parameters.

The degradation rate depends on the current density that flows through the diode and shows a cumulative effect, that is, only the time during which the device is submitted to the current flow counts for the degradation.

The degradations of OPVs are characterized by a loss of the conversion efficiency, which is indicated by a reduction of the area delimiting the J(V) characteristic of the cell under solar illumination. Both the open-circuit voltage V_{OC} and the short-circuit current density J_{SC} decrease, leading to a decrease of the fill factor FF. The lifetime for organic solar cells is defined using the metric T_{80}, which represents the time elapsed between the initial stabilized value (instant $T_0 = 0$) of the PCE and the instant T_{80} where 80% of the PCE is reached. Other ways to define the lifetime of OPVs can also apply, depending on the shape of the decay curve, which can show different evolution stages corresponding to different degradation mechanisms. In general, two stages of decay are observed in OPVs: a rapid initial decay of the PCE during a short time frame, followed by a slow decay, almost linearly, during a much longer time frame, which constitutes the real lifetime of the device (Turak 2013). For each decay stage, a set of two times (with initial instants T_0, T_{s0} and T_{80}, T_{s80} instants) are then defined to take the different degradation processes into account. The final failure occurring at the end of the second decay stage is generally rapid and corresponds to a complete degradation of the device. Because of the complexity of measuring the lifetime of different types of organic solar cells, as well as the multiple possibilities for failure processes, a global agreement for the stability testing protocols was proposed in 2011 and accepted by many research groups (Reese et al. 2011).

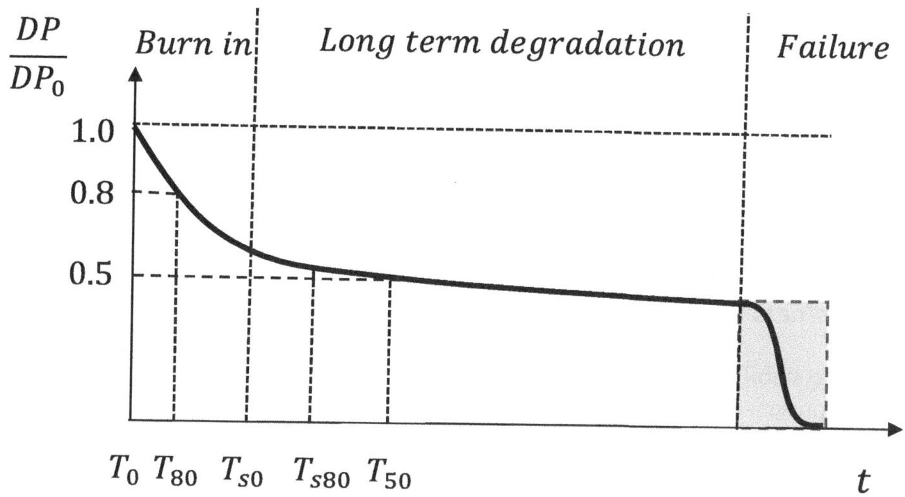

Figure 6.4. *Representative decay curve of the device parameter for defining the times used for describing the lifetime. The device parameter DP generally used for OLEDs is the luminance, for OPVs is the power conversion efficiency, and for OFETs is the saturation drain current. For a color version of this figure, see www.iste.co.uk/nguyen/defects.zip*

The degradations of OFETs are characterized by a loss of the transistor performance measured by plotting the on-state current, the transfer characteristic, as a function of operational time $I_{dsat}(t)$. The transistor lifetime is defined as the time T_{50} required for a loss of 50% of the initial saturated drain current (Han et al. 2006).

Figure 6.4 shows the time evolution of the device parameter (DP) and the definition of the different times used to determine the lifetime. The device parameter is the luminance (L) for OLEDs, the power conversion efficiency for OPVs and the saturation drain current I_{dsat} for OFETs, but other parameters are also used for this representation

It is worth noting that the determination of the lifetime of organic devices requires extended periods of performance measurements, from days to years, with well-defined protocols and experimental conditions. To simplify the investigations, *accelerated testing* is proposed as an alternative approach, in which the degradation of the studied device is artificially accelerated by increasing the stress parameters, such as the temperature, the applied voltage or the illumination intensity. One advantage of this technique is the possibility of lifetime evaluation of the device when the degradation follows an Arrhenius process, and the activation energy can be determined. By extrapolating the Arrhenius plot of the lifetime to the device use temperature, very large value of the failure times can be obtained with a simplified protocol. However, under accelerated testing measurements, the devices operate in extreme conditions and may experience degradation processes that would not occur in normal operation, leading to erroneous results. Furthermore, since the degradation of organic devices is complex and involves separate and/or mutual physical, chemical and morphological factors, accelerating tests (typically thermal, light or electrical stress) do not accurately reflect the degradation conditions and mechanisms in organic devices and are therefore controversial.

6.2.2. *Defects and degradation of organic semiconductor and devices*

As the stability of organic devices is linked to their performance, it is evident that the presence of defects affects the reliability through the degradation of the active material and device. The correlation between defects and degradation can be established by using the techniques previously applied for the proof of the defect/performance relationship. Here, the degradation can be evaluated from the change in the device parameter DP before and after modification of the defect concentration in the active semiconductor or in the whole device, with the defect formation process supposedly being known. Inversely, the defect measurements in the device before and after degradation make it possible to evaluate the defect parameters and hence, to understand the physical mechanisms of the failure process. It should be kept in mind that the described technique does not allow us to study the influence of a given and identified defect source on the global degradation of a device. It only provides information on the effects of the global defect states in the device on its degradation process. To separately investigate the defect sources and their effects, the technique should be applied to samples in which all the fabrication parameters are kept the same except for variation in the characteristics of a given defect source (for instance, different metal electrodes for interfacial defects).

6.2.2.1. *Defects and degradation of OLEDs*

OLEDs emit light by applying a voltage supplied by an external source at the terminals. Using high-quality organic materials and appropriate device structure, the turn-on voltage can be reduced to few volts. However, the applied electric field inside the device is large ($\geq 10^7\ V.m^{-1}$) because of the small thickness of the layers ($\leq 100\ nm$). As a high luminance requires a high current density, it can be expected that the charge carrier densities would have a major influence on the stability of OLEDs. A high current density in the device would produce high Joule heating, especially for large-area devices, which in turn accelerates other internal degradation reaction. Furthermore, in some cases, when high density of charge carriers occurs in the interfacial region, the excessive carriers that can produce an accumulation of charge in the vicinity of the recombination zone results in quenching of the created excitons and affects the performance and stability of the device.

The main sources of stress for OLEDs frequently encountered are moistures, oxygen, temperature and light (Scholz et al. 2015), although a large number of external and internal processes of degradation can be found and studied in these devices.

Early studies of degradation in OLEDs were focused on the formation of *dark spots*, which appeared on the cathode layer surface when the device was under electrical stress. Generally, the dark spots are observed in non-encapsulated diodes and progressively grow over time until complete extinction of the light emission. It has been established that these dark spots originate from areas of the organic layer containing defects and develop by the diffusion of moistures under the applied stress, until local peeling (*delamination*) of the cathode layer (So et al. 2010). By protecting the cathode layer with an efficient encapsulation, the dark spots formation is negligible and does not affect the luminance of the diodes. A second degradation process frequently observed in basic OLED structures is the *catastrophic failure*, which is a sudden complete loss of light output and a drastic voltage drop of the device during operation. This process is generally assigned to electrical short, which is due to the morphology of the untreated ITO substrate or/and the slow diffusion of ions from it into the layers of the device. A careful treatment of the anode surface before deposition of the organic layers allows this degradation mechanism to be minimized or suppressed.

The second source of stress for OLEDs is the high temperature at which the device is operating. The increase in temperature in a diode occurs either from the internal heating by Joule effect or from an external heating source (for instance, by exposure to sunlight or during the annealing process of the organic layers). It can reach values in the range of $70-90°C$ and may induce irreversible changes in the structure of the organic materials, leading to the alteration of the diode characteristics. The thermal behavior of organic materials depends on their *glass transition temperature* T_G, over which mechanical deformations of the films (especially, the film morphology changes and dewetting) can occur, as well as recrystallization processes. In addition, the high temperature is an accelerating factor for other physical processes occurring in the devices, including the different degradation mechanisms.

As a source of stress, light, especially UV, can interact with the organic semiconductor via the photochemical reaction and create keto groups (C=O) in the presence of oxygen in the material. These groups act as quenchers and degrade the operation of OLEDs. Another effect of light irradiation is the changes at the organic/metal cathode (or ITO anode) interface that lead to a deterioration in charge carrier injection, and hence degrade the diode operation.

Generally, the degradation of OLEDs under operation with constant current density manifests by a decrease in the luminance and a simultaneous increase in the applied voltage.

The luminance loss, which occurs during continuous operation of an OLED, is due to a decrease in the radiative recombination rate of excitons in the semiconductor layer. Several physical processes involving the defect states can be responsible for this decrease in luminescence. Indeed, traps can influence the effective mobility of the charge carriers, leading to a decrease of the charge balance factor, which in turn reduces the luminous efficiency (Meerheim et al. 2008). On the other hand, traps acting as non-radiative recombination centers and luminescence quenchers effectively reduce the photon emission. It has been demonstrated by experiments that the device degradation is related to the gradual generation of traps during its operation and the luminance decays are dependent on the trap concentrations (Kondakov et al. 2003).

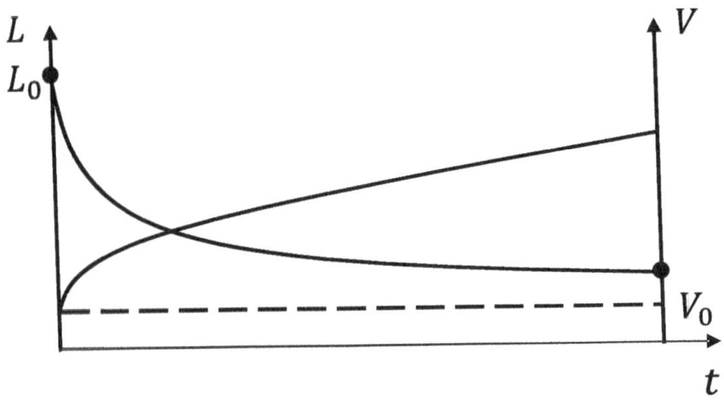

Figure 6.5. *Schematic evolution with time of the OLED luminance and applied voltage under constant current density conditions*

6.2.2.1.1. Small-molecule-based OLEDs

It should be noted that small molecule materials are usually deposited in thin films by evaporation under vacuum and the contamination by residual atmosphere is negligible. On the other hand, with appropriate substrate treatment, diffusion of metal ions from the ITO anode may be controlled or suppressed. For OLEDs using small molecules as an emitter material, several physical processes have been proposed to explain the degradation of devices in relation to the presence of defects.

In Alq3-based devices, the degradation is assigned to diffusion of indium ions released from the ITO anode into the emitter, which results in luminescence quenching (Lee et al. 1999). This explanation is supported by the enhancement of the lifetime observed in OLEDs using a surface-treated ITO or an HTL, which hampers the metal diffusion. Other investigations of the Alq3-based OLEDs based on the photoluminescence efficiency measurements suggest that the defects which are responsible for the device fluorescence decay are the degradation products of cationic Alq3 species of the organic emitter, which act as quenchers (Aziz et al. 1999). From the variation of the applied voltage with time in hole-only and electron-only diodes, the injection and transport of holes is found to be the main cause of the degradation.

However, the transport materials may also play a primary role in the degradation small-molecule-based OLEDs, as shown by a study of defects using TSC measurements in commonly used hole transport materials (HTMs), such as α-NPD or 1-NaphDATA (Steiger et al. 2001). The diodes of structure $ITO/HTM/Al$ were degraded by exposure to oxygen and humid atmospheres and the TSC measurements were performed for both degraded and non-degraded devices. The results showed that oxygen exposure does not affect the defects states in both HTMs and under humidity atmosphere, deep traps of activation energy ~0.5 eV are created in the band gap of 1-NaphDATA. No new defects are observed in the α-NPD layer.

A study analyzing the changes in both electrical and optical characteristics of OLEDs with emitting material of CBP doped with green phosphorescent emitter $Ir(ppy)_3$ allows us to clearly demonstrate the correlation between degradation and trapping processes (Schmidt et al. 2015). The complete structure of the OLED used comprises an ITO substrate, a hole injection layer HATCN, a hole transport layer (α-NPD), an emitter ($CPB/Ir(ppy)_3$), an electron transport layer (Bphen) and a cathode (Ca/Al). The accelerated aging of the diode is performed by applying a constant current to the device. Time-resolved electroluminescence spectroscopy (TRELS) was performed to obtain information on the excited state lifetime and showed a decrease in the lifetime after aging, without changing the emission spectrum. This variation is explained by the creation of quenching centers, which increases the non-radiative decay rate of the excited molecules, while the radiative rate remains constant after the device has aged. The location of the created defects was further determined by impedance spectroscopy and displacement current measurements by comparing the characteristics obtained in the diode before and after aging. The results indicate that the traps would originate from the emitter, precisely, in the vicinity of the emitter/electron transport layer interface. Other investigations in devices using Alq3 as an emitter have shown similar degradation causes, which are proposed as common mechanisms for a wide range of OLEDs (Kondakov et al. 2007).

The magneto-electroluminescence (MEL) method has been used to analyze the thermal degradation of Alq3-based devices of structure $ITO/PEDOT:PSS/NPB/Alq3/BCP/LiF/Al$ (Zhu et al. 2020). It has been shown that when the temperature of the diode is higher than the T_G of the emitter, the light emission is altered because Alq3 tends to aggregate and create structural change and traps. The traps limit the diffusion of excitons and, as shown by the MEL spectra, promote internal intersystem crossing transitions, which in turn decrease the radiative recombination in the organic semiconductor and degrade the device performance.

6.2.2.1.2. Polymer-based OLEDs

Polymer thin films are generally deposited from the solution by spin coating or printing techniques. The contamination possibility by the device processing is higher than that of small-molecule diodes, especially concerning oxygen and humidity.

Two conjugated polymer families are intensively studied for OLED applications: the PPV and derivatives, and the PF and derivatives.

Due to their synthesis and processing, PPVs films are generally contaminated by oxygen (Nguyen et al. 1993b), which act as trapping centers for injected charge carriers at the interface of the PPV and the contact layer. The defects affect the injection of charge carriers of both types and hence the influence of the charge transport and the device lifetime. Furthermore, creation of defects may occur in the volume of the polymer film and affect the transport properties and the diode lifetime in the same manner.

In earlier studies of PPV-based OLEDs, the correlation between the interfacial defects, the charge transport and the lifetime has been established for OLEDs of structure $ITO/MEHPPV/Ca/Al$ (Scott et al. 1996), in which the degradation occurs by two major modes. The first process is the oxygen diffusion from the ITO substrate into the polymer forming carbonyl moieties, which results in the quenching of the luminescence and the decay of the light output. The second process is the gradual formation of microscopic shorts, which reduce the active area and the intensity of the emitted light.

The role of chemical impurity defects has been put forward in a study of PPV derivative OC_8C_8 based OLEDs (Fleissner et al. 2009). It has been shown that halide defects strongly affect the lifetime by measuring the current–luminance–voltage characteristics over time in devices with various concentrations of residual bromine. As residual halogen impurities act as quenching centers (assigned to electron traps) in PPV, a high bromine concentration in the polymer leads to a fast luminance decay with time and a reduction of the device lifetime.

In diodes using a PPV copolymer, the Super Yellow PPV (SY-PPV), it has been demonstrated that the cause of device degradation is the formation of hole traps, which increases the non-radiative recombination of free electrons and trapped holes, leading to a decrease in the device luminance (Niu et al. 2018). The formation of the hole traps is explained by the interactions between excitons and polarons, which results in bond breaking and dissociated products, which act as hole traps.

In PF-based OLEDs, the most studied degradation process concerns the onset of a green-blue emission band, which occurs during device operation and changes the initial blue light emission. The degradation is attributed to charge carrier trapping by the keto defect (fluorenone) sites, which can be formed during synthesis and produces emissive recombination with photons of a lower energy than the initial emitting blue light (List et al. 2002).

Further investigations of OLEDs using PF-based LUMATION Green Series Light-Emitting Polymer (LEP) indicate the role of the PEDOT-PSS transport layer in the degradation of performance and lifetime ($\sim 1{,}500\ hrs$ at $1{,}000\ cd.m^2$) during operation (Kim et al. 2004). The cause of the degradation is attributed to the decrease in the hole-injection rate, which is related to the degradation of the interface transport layer/emitter by the acidic and hygroscopic nature of the PEDOT: PSS. It has been found that the ITO substrate is etched by PEDOT:PSS and diffusion of In ions to the polymer occurs, introducing impurity defects at the interface.

To understand the aging process in PF-based LUMATION Green Series OLEDs, the determination of trap states in devices before and after degradation by electrical stress has been investigated using the TSC method (Koden et al. 2007). To distinguish between electron and hole traps, three OLEDs structures have been used: $ITO/Ca/LEP/Ba/Al$ (electron-only device), $ITO/PEDOT{:}\,PSS/LEP/Au$ (hole-only device) and $ITO/PEDOT{:}\,PSS/LEP/Ba/Al$ (bipolar device). Analysis of the TSC spectra of fresh samples allows the determination of trap states in the bulk of the PF polymer, which include hole traps with an activation energy of $0.19\ eV$ and electron traps with an activation energy of $0.21\ eV$. On the other hand, the interface PEDOT:PSS/LEP contains trap states, including electron traps of activation energies with 0.16 and $0.23\ eV$ for electron traps and 0.24 and $0.40\ eV$ for hole traps. After the aging of the diode, the TSC spectra are recorded at a different degraded luminance ($L_{80} - L_{40}$) and analyzed. The bulk trap states are unchanged, whereas the interfacial trap states increase by degradation.

The Q-DLTS technique has been applied to determine the trap parameters of OLEDs using PF-N-Ph (poly(9,9-dihexylfluorene-co-N,N-di(9,9-dihexyl-2-

fluorenyl)-N-phenylamine), a polyfluorene derivative, as an emitter, before and after degradation by electrical stress (Nguyen et al. 2008). The structure of the bipolar diodes is $ITO/PEDOT{:}PSS/PF\text{-}N\text{-}Ph/Ca/Al$. In fresh samples, the Q-DLTS spectra show that five trap levels have been determined with energy activation from 0.14 to $0.58\,eV$ and with a density ranging from 10^{16} to $10^{17}cm^{-3}$. After degradation at half lifetime, no new trap levels have been found, but the density of some traps has increased. Shallow trap states remain stable, while deep traps show strong density variation, indicating a deterioration of material structure.

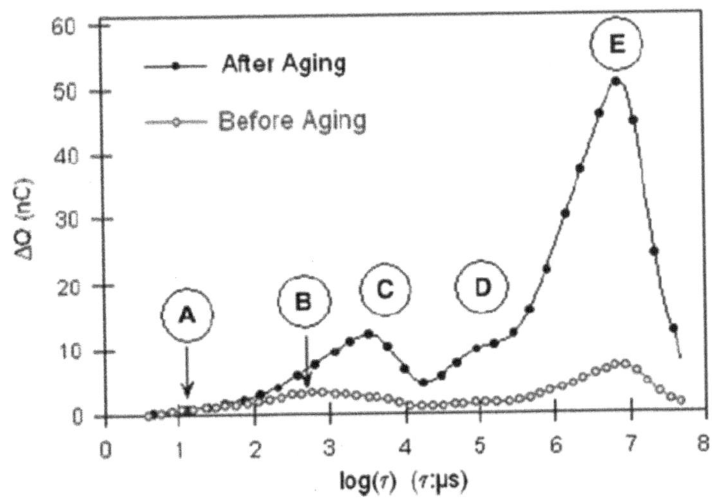

Figure 6.6. *Q-DLTS spectra of a fresh and an aged ITO/PEDOT:PSS/PF/Ca/Al diode at T = 300 K, with the following experimental conditions: charging time t_c = 1 s, charging voltage ΔV = +6 V (Nguyen et al. 2008, p. 92). For a color version of this figure, see www.iste.co.uk/nguyen/defects.zip*

6.2.2.1.3. Reliability of OLEDs

OLEDs can be used for either display or lighting applications. On the one hand, thanks to their high quantum efficiency, OLEDs enable high-quality and highly efficient displays, which have been commercialized as large size screens in mobile or fixed applications such as TVs or billboards. On the other hand, in the view of the prospect of energy saving in the future, OLEDs are promising for the lighting industry because of their low energy consumption when providing light of high luminance, and in the perspective of a share of 20% of world global energy consumption by lighting, OLEDs have an important role in the future electronics industry.

For displays, the required luminance is typically $L = 600\text{–}6{,}000\ cdm^2$, with an estimated lifetime of $T_{90} = 10{,}000\text{–}30{,}000\ h$, and for lighting, the luminance and lifetime should be $L = 1{,}000\text{–}3{,}000\ cd.m^2$ with a lifetime of $T_{70}{\sim}50{,}000\ h$.

The early OLEDs with a basic architecture and using small molecules or conjugated polymer emitters had a low efficiency. Gradually, it increased with the introduction of transport layers for electrons (ETL) and holes (HTL). However, according to the electron spin statistics, the emission from the radiative recombination of singlet excitons of fluorescent emitters is limited to 25% of the generated excitons. This limitation leads to an upper limit of the external quantum efficiency of ~5% in fluorescent devices. The use of phosphorescent emitters allows the emission efficiency to be increased by harvesting both the singlet and triplet excitons, thereby multiplying the theoretical fluorescence efficiency by 4. The efficiency of OLEDs has been greatly improved by associating fluorescence and phosphorescence processes in one emitting material, through the exploitation of excitons generated by the *thermally activated delayed fluorescence* (TADF). The principle of the generation of singlets and triplets upon electrical excitations is shown in Figure 6.7. It is based on the up-conversion of the triplet excitons to the emissive singlet level by *reverse intersystem crossing*, followed by a delay fluorescence to the ground state. This process requires a small singlet–triplet energy gap $({\sim}10 - 200\ meV)$ and results in a harvesting of both singlet and triplet excitons, making it possible to obtain an internal quantum efficiency of nearly 100% for the OLED (Adachi 2014).

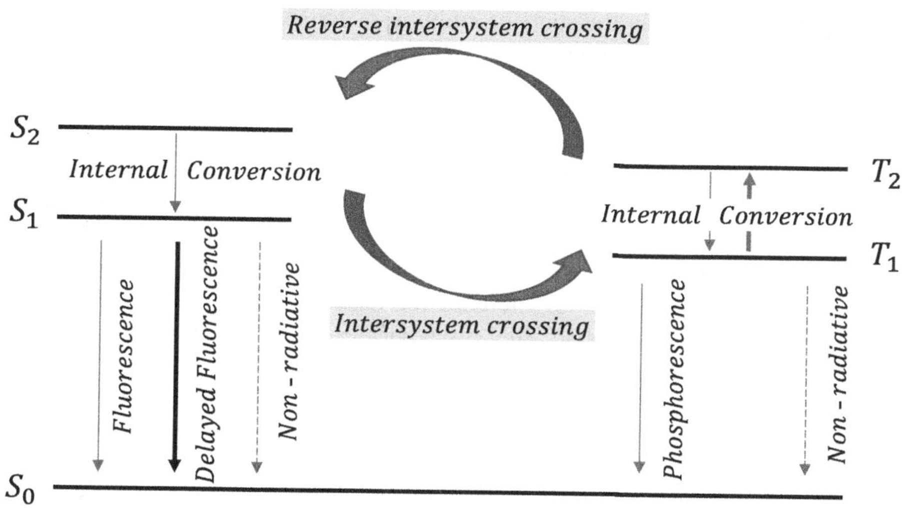

Figure 6.7. *Principle of thermally activated delayed fluorescence. For a color version of this figure, see www.iste.co.uk/nguyen/defects.zip*

In terms of practical efficiency, the *efficacy* of OLEDs represents the ratio of luminous flux to the power consumed and is expressed in lumens/watts (lm/W). For white light OLEDs, efficacy higher than 60 lm/W has been reported, which is the international standard for white light requirements. On the other hand, the driving voltage of OLED devices, which is an important parameter for the device operation, can be reduced to values as low as $\sim 3\ V$ (Murano et al. 2005), while yielding a luminance of 1,000 $cd.m^2$ with an efficacy of 17.5 lm/W. This is made possible by intentional doping of the transport layers. So, technical progress in OLEDs displays and lighting applications is obviously demonstrated, which can lead to a higher reliability of the light-emitting devices.

In terms of long-term stability, the lifetime of OLEDs increased by one order of magnitude every 5–10 years (Scholz et al. 2015). Improvement in the device lifetime has been achieved partly owing to a better understanding of the effects of defects on degradation processes of materials and devices, which further allowed us to find appropriate protection means to mitigate and slow down the operational degradation rate. Besides, the use of better quality and innovative materials for emitters has also improved the stability of the diodes. Most of the early OLEDs used small molecules such as Alq3, or conjugated polymers such as MEH-PPV as emissive layers. These materials have good physical properties and are still employed in OLED fabrication. Progress in the synthesis of organic materials then allowed the development of low operating voltage emitters (for instance, Ir(ppy)3) or phosphorescent emitters (for instance, FIrpic), which greatly improved OLED performance (Sasabe and Kido 2013). Similarly, organic materials with high thermal stability have been investigated to be used as transport or emitting materials in OLEDs, with a glass transition temperature often higher than 200°C, which can allow devices to operate in harsh environmental conditions (UV exposure and high operating temperature) without notable degradation (Lee et al. 2021).

Fluorescent devices have lifetimes (T_{50}) in the range of 10^5–$10^6\ h$ for green and red OLEDs, and $10^4\ h$ for blue OLEDs. Lifetimes of phosphorescent devices are in the range of 10^5–$10^6\ h$ for red OLEDs and 10^4–10^5 for green and blue OLEDs (Scholz et al. 2015). It is worth mentioning that these values are merely indicative, and many studies reported lower lifetimes for OLEDs using similar architectures and organic material emitters. The discrepancies often arise in the lifetime measurement conditions, namely the initial luminance of the devices. Even so, it can be stated that the OLED technology has become mature and OLED devices can be considered as reliable, since the required characteristics for both displays and lighting can be globally obtained for most of the available commercial applications, such as portable devices and TV screens, billboards, indoor luminaires. It should be noted that flexible OLEDs are not considered here, and this will be the case for all flexible organic devices. Indeed, the mechanical behavior of these particular components implies that each part of the device should be flexible and require a higher reliability compared to that of a rigid device. Although this can be feasible, the technology of flexible devices is still far from mature and needs more investigations to be considered as reliable.

6.2.2.2. *Defects and degradation of OPVs*

An organic solar cell converts photon power into electrical power when illuminated with solar light. From this working principle, a cell is continuously exposed to solar illumination during operation. This exposure may have two degradation effects. First, an increase in the temperature of the cell due to sun irradiation will occur. In a common silicon PV stand, the mean module temperature can reach 50°C (Faiman 2008), and it is reasonable to suppose a similar temperature for outdoor operating OPVs, although this does not consider the internal heat dissipation from the device. Because of this thermal degradation, the morphology of the layers may change by crystallization of the surface layer, and a phase transition may also occur in low glass transition temperature T_G materials. The second effect of the solar illumination is the photo-induced degradation of the organic layers, especially the active layer of the cell. The possible degradation effects are (i) photobleaching, which results in a reduction of the absorption intensity and polymer conjugation, creating additional structural defects, (ii) photooxidation of the polymer in presence of oxygen, leading to breaking of the conjugation and chain scissions and (iii) formation of radicals, which can act as deep traps for charge carriers and affect the efficiency and lifetime. In addition to these factors, the other intrinsic and extrinsic degradation sources will also be considered, as for all organic devices.

In real devices, depending on the use conditions of organic solar cells, several degradation processes can simultaneously occur, especially during outdoor experiments, and the identification of degradation processes is complicated. It is possible however, to distinguish between the degradation processes occurring in OPVs and obtain the impact of each stress factor by examining only the studied degradation factor and eliminating the effects of the others during indoor experiments. For instance, to study the influence of photo-degradation on the lifetime of OPVs, the device should be kept at a constant temperature (no thermal degradation and morphological changes) and under an inert atmosphere (no oxidative degradation) during the measurements. For indoor testing, the solar cell is usually illuminated under AM 1.5G white light ($100\ mW.cm^{-2}$).

The lifetime of an OPV is generally measured by the decrease over time of the open-circuit of the degradation parameter DP, which can be the open-circuit voltage V_{OC}, the short-circuit current density J_{CC} or the power conversion efficiency PCE. The decay curve of the device parameter DP comprises two main parts: the first part is a steep degradation indicated by a fast decrease of the parameter with time. This process is referred to as *burn-in loss*, whose origin is still unclear (Tamai et al. 2016). The burn-in step is necessary for the commercialization of electronic devices as it warrants a reliable product when delivered to the customers. However, the efficiency of the solar cells after a burn-in loss is of course lower than its initial value. The second part, referred to as a *long-term degradation*, is a slow degradation process, and its duration is much longer than that of the burn-in loss, and practically corresponds to the usable time of the cell. As the lifetimes of silicon-based solar cells can be estimated to be more than 20 years, and given that the mean duration of sunshine period is estimated to be ~1,000 h per year, a loss of 1% of the cell performance should be considered for practical applications of OPVs (Roesch et al. 2015).

It should be noted that in general, the decays over time of the different degradation parameters DP are not identical for a solar cell. Depending on the physical degradation process to be studied, the decay curve of a DP parameter can be chosen and analyzed and, in some cases, there can be a correlation between the decay curves with different DP parameters.

The correlation between defect formation and the lifetime of OPVs can be established by determining the defect parameters in a cell before and after degradation. Then, the comparison of these parameters allows for validation (or not) of the relationship lifetime defects. Moreover, by precisely defining the degradation conditions (light, temperature, environmental media), information on the defect formation process can be obtained and its influence on the device lifetime can be analyzed and discussed.

6.2.2.2.1. Small molecule (SM)-based solar cells

There are two types of device structures for small-molecule-based solar cells: the planar donor–acceptor (D–A) bilayer and the bulk D–A heterojunction (BHJ). The most frequently used materials for donors are phthalocyanines (CuPc, ZnPc), oligothiophenes, polyacenes (including anthracene, tetracene, pentacene, rubrene) and perylene diimides (PDI). Materials used for acceptors are fullerenes, including PCBM, and non-fullerene acceptors (ITIC polymer and derivatives). When only SMs are used, the BHJ mixed layer is obtained by co-deposition under vacuum of the SM materials. Because of the possible excitons quenching at the device active layer/cathode interface, a thin layer of large band gap is deposited between these layers, acting as an *exciton blocking layer* to prevent excitons from quenching. The small molecule materials usually employed for the exciton layer are BCP and BPhen.

To study the effects of oxygen and water on degradation, the lifetime of SM solar cells with an active mixed layer composed of zinc phthalocyanines (ZnPc) and fullerene (C_{60}) was studied by measuring the characteristic parameters over time in devices of structure $ITO/MeO\text{-}TPD{:}\,C_{60}F_{36}/ZnPc{:}\,C_{60}/C_{60}/BPhen/Al$ (Hermenau et al. 2011). Under AM 1.5G illumination, different behaviors were observed, depending on the medium to which the samples were exposed. The device lifetime decreased in the following order: nitrogen > oxygen > water. Analysis of the samples by the time-of-flight secondary ion mass spectrometry (TOF-SIMS) and X-ray photoelectron spectroscopy (XPS) techniques suggests that both ZnPc and MeO-TPD are contaminated by oxygen and water, and that water and oxygen diffuse to the active layer through the grain boundaries and the pinholes in the aluminum electrode, respectively.

The role of defects in the burn-in loss was investigated in devices using active layers composed of SubPc and fullerenes C_{60} (C_{70}) (Tong et al. 2013). Degradation of devices of structure $TO/MoO_3/SubPc{:}\,C_{60(70)}/PTCBI/Al$, where PTCBI is used as an exciton blocking layer, was carried out under AM 1.5G irradiation in the absence of oxygen and water. The burn-in loss in the PCE was saturated after 10 h of exposure in planar devices and 48 h in mixed devices. Analysis of the variation of the external quantum efficiency overtime of the devices before and after degradation indicates that the fullerene layer C_{60} has strongly changed. The degradation process is assigned to the reduction of excitons generated in fullerene, which is due to the *exciton-induced trap* formation. This defect formation process occurs by the chemical reactions between highly energetic excitons with the molecular materials absorbing the photons. These traps act as deep charge recombination centers in the photoactive layer and reduce the efficiency of the charge collection. The saturated trap density responsible for the aging degradation in the active layer was estimated to be $1.2 \times 10^{17} cm^3$.

6.2.2.2.2. Polymer-based BHJ solar cells

The active materials used in polymer-based BHJ solar cells are usually formed by blending a conjugated polymer belonging to the PPV or PT family (as a donor) and a fullerene, mainly with PCBM (as an acceptor). The most studied blend is the P3HT:PCBM because of their good physical properties and relative high photochemical stability. However, in blend configuration, other degradation pathways may emerge for some mixtures of P3HT:PCBM derivative and lead to a premature device failure.

Investigations of the lifetime of PPV-based BHJ solar cells (unencapsulated) of structure $ITO/PEDOT{:}\,PSS/MDMO{:}\,PCBM/Al$ were performed for different ambient conditions (under light, in the dark, under air, dry oxygen and humid nitrogen) (Kawano et al. 2006). The measured device half-life was ~3 h in ambient air under both light and dark conditions and ~2 h in humid nitrogen atmosphere. The degradation was attributed to a water-induced process, which originates from the PEDOT:PSS layer. The study of the photodegradation mechanism in these devices was performed under continuous 1.5 AM illumination and in defined environmental conditions (Pacios et al. 2006). The analysis of degraded devices shows a reduction of optical absorbance of the polymer with scission of the backbone and loss of the conjugation length. In addition, measurement of the absorption kinetics by the *transient absorption spectroscopy* (TAS) technique indicated an increase of the energetic tail of trap states after degradation, which was assigned to the changes in conjugation length and the presence of chemical defects. The use of Monte Carlo simulations allowed for the interpretation of the degradation mechanism by the formation of hole deep-traps in the polymer.

Another degradation study of polymer-fullerene-based solar cells was performed using devices of structure $ITO/PEDOT{:}\,PSS/PC_{60(70)}BM{:}\,PCDTBT(P3HT)/Ca/Al$, where PCDTBT and P3HT are conjugated polymers (Heumueller et al. 2015). The loss of the open-circuit voltage V_{OC} was measured over time with exposure of the device to AM 1.5G radiation, under vacuum and at constant temperature for three days. Using transient photovoltage and charge extraction methods, the recombination dynamic charge carrier densities of fresh and degraded solar were compared and related to the V_{OC} voltage. The observed burn-in loss of the devices could be described by an increase of the energetic disorder of the material DOS, which did not change the recombination dynamics of carriers, but changed their distribution, leading to broad density of states and therefore, a reduction of V_{OC}.

Among the blends using a polythiophene derivative as the photoactive material, the mixture of P3HT and PCBM has been intensively studied for their good stability. The failure mechanisms of these devices have been investigated by numerous works, which confirmed the formation of defects in the blends through the processes explained previously. These are: (i) photochemical scission of the polymer chains involving oxygen and light, and (ii) metal ion diffusion and reaction with the polymer. A further process has been also proposed to explain the degradation of the blends, that is, phase separation, according to which under favorable conditions, one of the two components of the blend tends to diffuse and agglomerate. This phase separation usually occurs near the interface between the blend and a contact layer, and affects the charge separation by reducing the density of the D/A contacts. For the P3HT:PCBM-based solar cells, some interesting studies on the correlation between the lifetime and the defect formation are given below.

Devices of inverted structure $ITO/ZnO/P3HT{:}\,PCBM/PEDOT{:}\,PSS/Ag$ were degraded by continuous exposure under AM 1.5G illumination (Norrman et al. 2010). The evolution of solar cell parameters was recorded as a function of time with encapsulated and unencapsulated devices, which were placed in different atmospheric conditions (dry oxygen, humidity, in the dark and under illumination) for comparison. The lifetime of unencapsulated cells is short (for instance, T80 ~1.9 h in water-containing atmosphere), compared to that of encapsulated ones (T80 ~360 h).

By combining XPS and TOF-SIMS (time-of-flight secondary ion mass spectroscopy) techniques, it was found that P3HT:PCBM and PEDOT:PSS are contaminated by oxygen in the dark (at low rate) or under illumination (at higher rate). For both materials, a phase separation occurs at the PEDOT:PSS/P3HT:PCBM interface and the P3HT:PCBM/ZnO interface.

Only a few experiments have been performed to quantitatively determine the trap parameters in degraded BHJ solar cells. For instance, TSC experiments were used to study the photodegradation of encapsulated P3HT:PCBM-based solar cells of structure $ITO/PEDOT:PSS/P3HT:PCBM/Ca/Al$ (Kawano and Adachi 2009). In these devices, the cell characteristics were gradually degraded with the exposure time under AM 1.5G illumination, but were recovered upon annealing up to 150°C. The TSC spectra recorded in a degraded cell indicate three peaks of activation energies $E_{t1} = 0.71\ eV$, $E_{t2} = 0.81\ eV$ and $E_{t3} = 0.91\ eV$, which are assigned to defects at the PEDOT:PSS/P3HT:PCBM interface, the P3HT:PCBM/Ca interface and the ITO/P3HT:PCBM interface. The degradation process in these devices is supposed to originate from the accumulation of the charge carriers in the trap sites at the interfacial regions: holes at the anode and electrons at the cathode. The recovery of the cell characteristics after annealing would be due to the release of trapped carriers from the defect centers.

Degradation in P3HT:PCBM devices has been investigated using the Q-DLTS technique to observe the changes in trap parameters in solar cells before and after degradation by exposure to AM 1.5G illumination (Nguyen et al. 2012). To correctly interpret the measurement results, the trap parameters are first determined in freshly encapsulated devices of structure $ITO/PEDOT:PSS/P3HT/Ca/Al$ and $ITO/PEDOT:PSS/P3HT:PCBM/Ca/Al$, in order to distinguish between the traps in the polymer and those induced by the fullerene. In P3HT devices, five trap levels of activation energy in the range 0.9 to 0.47 eV are determined. Upon incorporation of PCBM to P3HT, a new deep trap level is found with a high relaxation time, but its activation energy could not be determined with the experimental conditions used. Moreover, an increase in density of the existing traps in the polymer is observed, indicative of a higher disorder in the polymer structure. After exposure of the P3HT:PCBM devices to the light simulator for 20 h, the Q-DLTS spectra of the cells show a strong decrease in intensity of the existing traps in the polymer, but no new trap level is observed. The reduction of the number of defects can be assigned to a diminution of disorder in the blend upon aging, indicating a rearrangement of the donor and acceptor domains and possibly, a phase separation between P3HT and PCBM. Aggregation of the fullerene in the blends reduces the long-term stability of the solar cells.

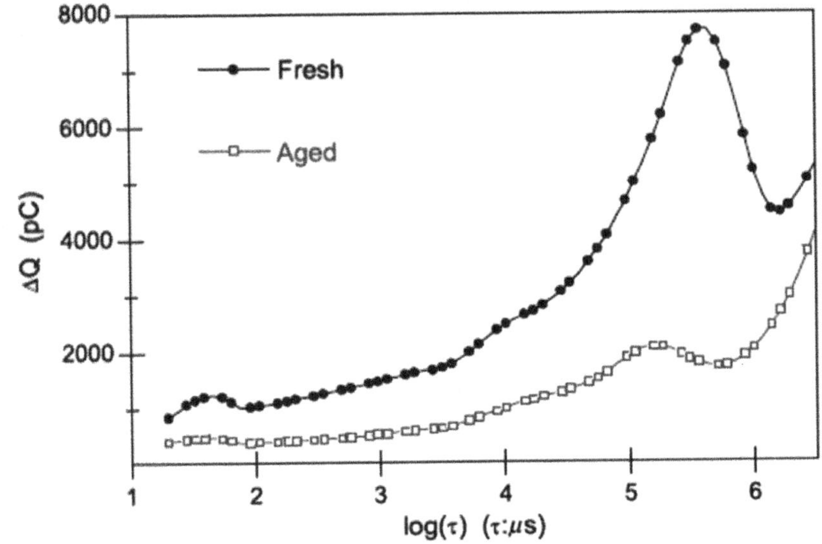

Figure 6.8. *Q-DLTS spectra of a fresh and an aged ITO/PEDOT/P3HT:PCBM/Ca/Al solar cell at T = 300 K, with the following experimental conditions: charging time t_c = 1 s, charging voltage ΔV = +3 V (Nguyen et al. 2012). For a color version of this figure, see www.iste.co.uk/nguyen/defects.zip*

Although fullerene acceptors have been widely used with conjugated polymer for OPV active materials with high electron mobility, their absorption of the solar spectrum is weak, and their band gap is not easily modified because of the rigid structure of the fullerenes. Therefore, the search for new organic acceptor materials containing no fullerene or non-fullerene acceptors (NFA) has been actively developed (Cheng et al. 2018). These materials have some advantages over the fullerene acceptors for the possibility to tune the band gap, the morphology and also the electron energy levels. The most promising NFAs are the ITIC (Lin et al. 2015), IDTBR (Holliday et al. 2016) and derivatives, which are organic compounds of fusing ring-based and push–pull structure A–D–A. ITIC has a strong absorption in the visible spectrum, a low LUMO and a high HOMO level, and devices using the PTB7-TH:ITIC blend as an absorber yield a mean PCE of ~6.5 %, which is close to that of PTB7-TH:PCBM-based cells. The degradation mechanism of $ITIC$-based devices under AM 1.5 G illumination was studied using encapsulated structures of $ITO/ZnO/PTB7-TH{:}\,ITIC\ MoO_3/Ag$ (Park and Son 2019). A decrease of the device PCE at T70 was observed after 50 h and relates to burn-in losses. By making use of NMR and FTIR techniques to analyze the degraded ITIC films, it was shown that under UV radiation, the vinyl group of the structure at the ZnO/ITIC interface is broken and provides trap sites to enhance the charge carrier recombination and reduce the efficiency of the cells.

For comparison, investigations of the burn-in loss in BHJ solar cells containing NFAs of the IDBTR and ITIC families blended with several donor-conjugated polymers such as PTB7-TH and PBDB-T have been performed by photostability tests under continuous AM 1.5G illumination for 120 h (Clarke et al. 2021). The structure of the encapsulated devices is $ITO/ZnO/active\ layer/Ag$. Depending on the blend composition and mainly on the choice of NFA, different stability behaviors have been observed. After degradation, the charge carrier density was found to be unchanged in devices using an IDTBR derivative and increased in devices using ITIC derivatives. This increase was attributed to the formation of sub-band tail states, which act as additional traps, and reduce the charge mobility. Optical spectroscopy (UV, PL and Raman) was then used to analyze the structure of the blend films before and after photodegradation. While there have only been negligible changes in the spectra of IDBTR blends after aging, notable modifications in those of ITIC blends have been observed. The degradation mechanism in the ITIC acceptors is assigned to the breaking of the chemical bonds by photochemical reactions, creating unbonded end-groups which act as trapping centers, leading to poor stability of the solar cells.

6.2.2.2.3. Perovskite-based solar cells

The basic perovskite $MAPbI_3$ films in solar cells can be obtained by deposition, either from the solution of PbI_2 and CH_3NH_3I in one or two stages, or by co-evaporation under vacuum. As for conjugated polymers, using the wet deposition process may introduce chemical impurities and by-products acting as defects affecting the lifetime of the devices. Several factors can favor defect formation in perovskites.

Extrinsic factors such as humidity and oxygen in environmental media affect the material by chemical reactions with its structure, with or without light, creating defects and disorder, as already seen in small molecules and polymers. Using appropriate encapsulations, the effects of these degradation factors can be limited or suppressed.

Intrinsic factors include (i) moisture and oxygen from the material, (ii) heat and (iii) ion migration. The two first factors also operate in other types of OPVs, and the last factor is more specifically characteristic of perovskite-based cells, due to their chemical structure. Indeed, perovskite has the ABX_3 structure in which vacancies and interstitials are likely present because of their low formation energies. In basic perovskites, the most probable defects are (V_I), (V_{MA}) and (V_{Pb}) vacancies, and iodide interstitials (I_I) that can move between the lattice sites by hopping under an applied electric field. Under light and heat conditions, or a combination of both, ion migration is accelerated and causes changes in the structure of the materials, decomposing the perovskite and resulting in phase segregation (Kamat and Kuno 2021).

For the influence of moisture and oxygen on the degradation of perovskite cells, the lifetime of $MAPbI_3$-based solar cells under continuous AM 1.5G illumination in air atmosphere has been studied in encapsulated (with glass) and unencapsulated devices in order to understand the effects of environmental media on the degradation process (Kim et al. 2016). The structure of the studied devices is $FTO/blocking\ TiO_2/mesoporous\ TiO_2/MAPbI_3/spiro\text{-}OMeTAD/Ag(Au)$. In unencapsulated cells, the PCE is reduced by one half (T50) after ~ 4 h of illumination in ambient air. In the same exposure conditions, encapsulated cells show a longer lifetime of $T50 \sim 150\ h$, but the PCE continuously decreases, indicating an intrinsic degradation process. The degradation by light exposure of the unencapsulated cells is assigned to the decomposition of the perovskite films induced by light and moisture, favored by the contact with the spiro-OMeTAD layer. By using impedance measurements, it has been found that a new phase has formed in this interface, which gradually introduces a barrier layer and hinders charge separation. In the encapsulated cells, the formation of the barrier is slower, but still operates, leading to a decrease in PCE of the devices. Replacement of spiro-OMeTAD layer by a NiO layer has been proven to suppress the interfacial layer and improve the device lifetime.

For the thermal degradation of perovskite, the stability of $MAPbI_3$-based solar cells under one sun illumination was investigated under different environmental conditions: (i) low temperature (-20°C) with no humidity, (ii) high temperature (55/85°C) with low relative humidity (10%) and (iii) high temperature (55/85°C) with high relative humidity (80%) (Han et al. 2015). Here, the temperature of 85°C is that reached during normal operation in full sunlight of the cells. The structure of the encapsulated and unencapsulated solar cells is $FTO/TiO_2/MAPbI_3/spiro\text{-}OMeTAD/Ag/Pt$. The lifetime (T50) is longer than 120 h for the cells in conditions (i), ~80 h for the cells in conditions (ii) and ~4 h for the cells in conditions (iii). Analysis of the degraded perovskite devices and materials by SEM and XRD techniques indicates a decomposition of the perovskite into PbI_2, the process being accelerated by heat in the case of encapsulated cells.

The thermal decomposition of $MAPbI_3$ can occur by two possible pathways according to

$$MAPbI_3 \leftrightarrow CH_3NH_2 + PbI_2 + HI \qquad [6.11]$$

$$MAPbI_3 \rightarrow CH_3I + PbI_2 + NH_3 \qquad [6.12]$$

Reaction [6.11] is reversible and can occur at relatively high temperatures (>85°C). It is then plausible to suppose such a decomposition process in solar cell degradation. Reaction [6.12] is irreversible and can only occur at very high temperatures (>150°C). In addition to a decrease of the active site density, the thermal degradation produces byproducts, which further react with the initial material enhancing the degradation.

When a perovskite-based solar cell is operating at high temperatures, in addition to a potential decomposition of the active layer, the diffusion of ionic species in the material may increase, leading to a decrease in cell performance. An example of thermally activated ion diffusion has been shown in solar cells of structure $FTO/mesoporous\ TiO_2/MAPbI_3/spiro\text{-}OMeTAD/Au$) (Domanski et al. 2016). Comparison of the PCE of devices aged at 20°C and 75°C for 15 h under AM 1.5G illumination in nitrogen atmosphere shows that a large loss occurs after only ~2 h in the device aged at high temperature, while in the device aged at 20°C, the loss is negligible. The structure of the perovskite examined by optical absorption and XRD spectroscopy shows no notable difference between the studied cells. On the contrary, TOF-SIMS elemental depth profiling clearly shows that Au ions diffuse across the spiro-OMeTAD layer to the inside of the perovskite layer and accumulate near the $TiO_2/MAPbI_3$ interface. Other investigations of the effects of long-term light exposure and electrical bias on encapsulated solar cells of structure $FTO/blocking\ TiO_2/mesoporous\ TiO_2/MAPbI_3/spiro\text{-}OMeTAD/Au$ also revealed a diffusion of gold and iodine in the devices (Cacovich et al. 2017). Here, the cells were exposed to an AM 1.5G irradiation for 200 h, using only the visible spectrum in order to avoid degradation due to UV radiation. The decay of the PCE to a half of its initial value ($T50$) was observed at about 20 h, while the measured current–voltage $J(V)$ characteristics indicated an activation of the trap states in the perovskite, as well as the bimolecular recombination of the device induced by the light stress. Using scanning transmission electron microscopy in conjunction with energy-dispersive X-ray spectroscopy (STEM-EDX), the diffusion of Au and iodine is identified. While iodine accumulates at the spiro-OMeTAD/Au interface, gold is found in both the active perovskite and the TiO_2 layers, under clusters at some specific regions. Degradation of the PCE was attributed to the diffusion of gold from the electrode and iodine from perovskite to the interfacial regions of the device, which altered the charge carrier transport and collection.

It should be noted that materials of the charge transport layers commonly used in perovskite-based solar cells to achieve high efficiencies may also constitute additional sources of defects of the devices. For instance, the hole transport layer of Li-spiro-OMeTAD is highly hygroscopic and when exposed to air at $T = 80°C$, is able to introduce moisture to the perovskite films on which they are deposited (Habisreutinger et al. 2014) and alter the structure of the active layer. For electron transport layers, metal oxides such as TiO_2 or SnO_2 are prone to provide excitons under UV radiation, which further interact with available oxygen to create highly reactive superoxide O_2^+ (process known as *photocatalysis*). In perovskite-based cells, these defects tend to oxidize and decompose the absorber, becoming the active sites for the onset of degradation (Aristidou et al. 2015).

Despite numerous investigations of the defect states and the degradation processes of perovskite-based solar cells, the correlation between them has not been clearly understood and established. The two main reasons for this are the complexity of the structural defects in perovskite materials, which can be simultaneously involved by a specific tress applied to the device, and the complexity of the degradation processes in these materials, which may interact and mask the effects produced by each of the degradation factors.

6.2.2.2.4. Reliability of organic solar cells

Similar to OLEDs, the degradation processes of OPVs are mainly caused by materials, interfaces between layers and device architectures, and are accelerated by environmental stresses. Often, these processes are intimately linked to the presence of defects, which are either already formed in materials or interfaces before degradation, or created and developed in materials or interfaces as a consequence of the process.

The progress in performance and lifetime of OPVs is closely dependent on the evolution of the active organic semiconductor used as the absorber.

For the performance of the organic solar cells, in the first stage of the OPV developments, the most studied materials for absorber were the P3HT:fullerene acceptor blends used in cells of BHJ structure. With optimal concentrations of materials, the efficiency of P3HT:PCBM-based solar cells remains rather weak (~5%) (Cheng et al. 2013). Furthermore, the cells suffer from structural defects of P3HT, blend instability with possible phase separation and poor light absorption of the fullerenes. New donor–acceptor blend materials have then been studied and developed to improve the performance of organic solar cells. Non-fullerene acceptor (NFA) materials have especially been synthesized with highly tunable properties and various structures, and used with new donor polymers based on the benzodithiophene (BDT) unit, to form blends of D–A or A–D–A type in OPV devices. The efficiency of the cells using these blends, for instance, with BTP-eC9 and PBQ_xF as absorbers, can be higher than 18% (Cui et al. 2021), approaching the performance of some perovskite-based solar cells. Indeed, the PCE of the early perovskite devices using the organometal halide $CH_3NH_3PbX_3$ with X = Br or *I* as a semiconductor sensitizer in dye-sensitizer solar cells, was modest (3-4%) (Kojima et al. 2009). To improve the efficiency of perovskite solar cells, new compositions have been studied, and it has been shown that the formamidine lead triiodide ($FAPbI_3$) has interesting optical (narrow band gap, high absorbance) and transport (long diffusion length) properties that can favor and enhance the energy conversion processes of the cells. Using an absorber of $FAPbI_3$ passivated by 2% anion formate ($HCOO^-$), it is possible to obtain a PCE higher than 25% by neutralizing the halide vacancy defects (Jeong et al. 2021). Taking the reference efficiency of 25% for silicon-based cells (Battaglia et al. 2016), we can see that the conversion performance organic solar cells is becoming quite close and comparable to that of standard commercialized cells.

On the stability of the organic solar cells, the lifetime of the early BHJ solar cells is rather short. For instance, the reported value for inverted and nonencapsulated P3HT:PCBM solar cells is $T_{80}\sim 650\,h$ (Savva et al. 2015) The cell lifetime is significantly improved when using non-fullerene acceptors and new polymer donor materials to replace the P3HT:PCBM blends. When using the PCE-10:BT-CIC blend as an absorber in cells with protective buffer layers at each side of the BHJ and a UV-filter deposited on the glass substrate, the initial efficiency of the cell exposed to AM 1.5G irradiation decreased to 94% after 1,900 h at 55°C. The extrapolated lifetime of the studied cells is determined to be $T_{80} > 56{,}000\,h$, that is, more than 30 years by supposing an average of 5 $kWh.m^2$ of sunlight per day (Li et al. 2021). A similar trend of the lifetime evolution of perovskite-based solar cells has been observed with the use of novel or modified absorber materials, which are more stable than initial perovskite compounds ($CH_3NH_3PbI_3$ and $CH_3NH_3PbBr_3$) (Cui et al. 2021). Depending on how the devices are structured and measured, the extracted lifetimes vary. It should be noted that most of the stability investigations are focused on one factor or aspect of the possible degradation mechanisms, and report on the device lifetime after improvement by the modifications of the cells (materials including the absorber and transport layers, processing conditions, architecture and encapsulation). The testing conditions of the devices are often different, depending on the degradation factor to be examined. Therefore, the reported lifetimes from these studies show large discrepancy between different perovskite-based solar cells. They can vary from a few minutes for early perovskite (MAPI) cells to thousands of hours for improved stability cells (Grancini et al. 2017). There is unfortunately no report on the lifetime of a prototype of perovskite-based solar cells, which could be used as a benchmark to compare the stability of devices. For OPVs, stability assessment protocols (ISOS) have already been established (Reese et al. 2011). These protocols define the measurement conditions of solar cells, including dark, outdoor, temperature and humidity cycling with different testing levels (basic, intermediate and advanced). Therefore, despite the improvement of the device lifetime obtained by diverse engineering techniques on the structure design, materials, processing and encapsulation, the current stability of perovskite cells is still not good enough for practical applications for the energy conversion by considering both performance and lifetime factors. In 2020, a consensus statement on procedures for testing the stability of perovskite solar cells has been proposed to unify the stability assessment and understand the failure modes of these devices (Khenkin et al. 2020).

The overall performance of BHJ and perovskite OPVs reasonably makes it possible to consider these devices as reliable for certain practical applications (of low power) in the conversion of solar energy to electricity. However, for replacing current silicon-based solar cells, their lifetime should be greatly improved, especially the perovskite cells for which the stability study protocols are still missing.

As for flexible solar cells, the observation made on OLEDs can be applied to BHJ blend and perovskite-based devices. In addition to the stress sources encountered in rigid devices, the mechanical stress due to the flexibility (tensile strain by bending or stretching) can generate structural defects, which in turn accelerate the degradation mechanisms. Furthermore, as all parts of the devices should be flexible, the reliability of each part should be insured, with particular attention paid to the electrodes, the substrates and the encapsulation layers. As the properties of the current flexible materials are not sufficiently good for replacing those used for rigid devices, the flexible devices are not yet reliable, contrary to expectations, as long as the problems of stability are not solved.

6.2.2.3. *Defects and degradation of OFET*

Except for certain specific structures such as OLETs (organic light-emitting diodes) or flexible transistors deposited on transparent substrates, an OFET does not need a transparent conducting oxide electrode like an OLED or an OPV for its operation. This difference ensures that OFET devices are not permanently exposed to light during operation, and therefore limits the effects of light and, to a lesser extent, heat on the device stability. The main physical process influencing the stability of OFETs is the electrical transport of charge carriers in the active material, and the degradation of OFET devices is generally described through electrical characteristics that are used to evaluate their performance. These characteristics are the carrier mobility μ, the threshold voltage V_T, the subthreshold swing S and the on/off current ratio I_{on}/I_{off}. The correlation between the defects and degradation in OFETs is studied by determining the variation of the chosen decay parameter and its defect states (when possible) over time. In order to investigate the physical process of degradation occurring in the devices, specific defects can be intentionally introduced to the structure, and studying their effects on degradation will allow us to understand and distinguish between different degradation factors in competition, as already seen in OLEDs or in OPVs.

6.2.2.3.1. Effects of bias stress on OFET degradation

Under a continuously applied gate voltage V_{GS}, the performance of an OFETs is degraded, as commonly observed by a shift of the threshold voltage V_T, and/or an increase in the subthreshold swing S, and/or a decrease in the charge carrier mobility μ_{FET} and/or a hysteresis increase in the transfer characteristics. The effect is generally assigned to the trapping of charge carriers in the channel, and one of the principal characteristics is the slow trapping rate that can spread over seconds, days or months. If the charges circulating in the channel between the source and the drain are trapped, they no longer contribute to the current flow. and the dielectric/semiconductor interface being charged will hamper the transport of the free carriers and consequently degrade the performance of the device. Under this condition, to maintain the current density in the channel, an increase in the applied gate voltage V_{GS} will be necessary, which will produce a shift of the threshold voltage V_T. Therefore, observation of the V_T voltage shift generally suggests a charge trapping process at the dielectric/semiconductor interface. Another possible consequence of the charge trapping in the channel is the change of the shape of the transfer characteristics. If the traps in the interface region are shallow, this shape will change upon trapping, leading to a decrease in the carrier mobility. If the traps are deep, this shape is unchanged and the charge carrier mobility will not be affected.

The defects that are responsible for the charge carrier trapping by bias stress are localized in the channel vicinity in different possible regions: in the semiconductor, in the dielectric and in the semiconductor/dielectric or semiconductor/electrode interfaces. The localization of the defects in devices is not clearly established and seemingly depends on the nature and quality of the materials and the processing. In solution-processed organic films, the formation of grain boundaries in the semiconductor has been found to affect the OFET stability under bias stress by providing trapping sites which induced the device degradation (Nguyen et al. 2016). In other studied devices, the nature of the dielectric layer is found to strongly influence the bias stress instability, and treating or modifying it without changing the semiconductor can notably improve the electrical characteristics of the transistors (Liu et al. 2016). The contact region between the different layers is also known to have high defect densities and impacts on the transistor stability, as proven by the study of OFETs with different metal electrodes for the source and the drain (Yan et al. 2011).

As an example, the aging effects of OFETs of structure $Si/SiO_2/PMMA/pentacene/Au$ (source and drain) have been studied by applying bias stress with positive and negative voltages to the gate electrode for a period of 48 h (Cipolloni et al. 2007). The degradation was measured by the variations of the threshold voltage V_T, which shifts in the direction (positive or negative) of the gate voltage applied during stress (Figure 6.9).

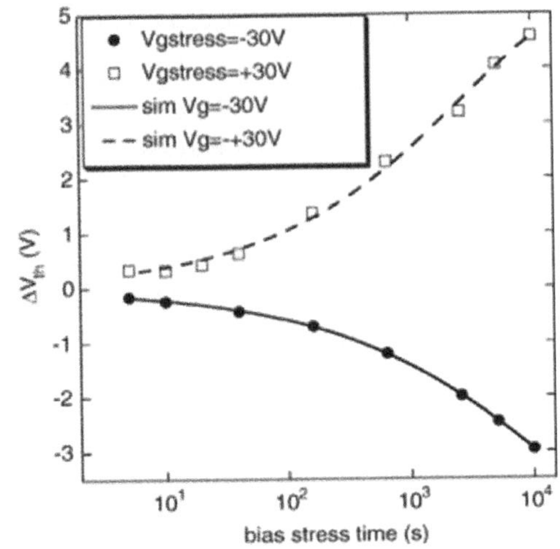

Figure 6.9. *Threshold voltage variations for pentacene OFETs during bias stress with fitted stretched exponential functions. (Cippolloni et al. 2007, p. 7546)*

The V_T voltage was determined by extrapolating the linear part of the transfer characteristics $I_D(V_{GS})$ to the gate voltage axis, and its variation is calculated by $\Delta V_T(t) = V_T(t) - V_T(0)$. The threshold voltage variation can be fitted by a stretch exponential function of the form:

$$\Delta V_T(t) = \Delta V_\infty \left\{1 - exp\left[\left(\frac{t}{\tau}\right)^\beta\right]\right\} \quad [6.13]$$

where ΔV_∞ is the maximum threshold voltage variation. The relaxation time is thermally activated and can be expressed as:

$$\tau = \nu^{-1} exp\left(\frac{E_t}{kT}\right) \quad [6.14]$$

where E_t is mean trap energy level and ν^{-1} is a frequency factor.

After degradation, the device was kept in the vacuum for 48 h and the transfer characteristic was completely recovered for positive bias stress and partially recovered for negative bias stress. This recovery suggests that charge carriers are first trapped in defect sites and then gradually released. The effect of bias stress is analyzed by a trap state model, assuming a uniform distribution of localized states DOS, including defects from grain boundaries and the PMMA/pentacene interface. The defect density includes tail states of energy E_t and density N_t and deep states of energy E_d and density N_d. The voltage variations are related to the changes of deep localized states, which increase in density during the positive bias stress. It should be noted that several other physical processes have been proposed to explain the bias stress effect observed in OFETs and the localization of the trapping centers (in the dielectric, in the dielectric/organic semiconductor interface or in the semiconductor) is also under debate.

6.2.2.3.2. Effects of environmental stress on OFET degradation

As in OLEDs or in OPVs, environmental stresses such as oxygen, humidity, heat or light affect the operation of OFETs and cause their degradation. It should be noted that very often, an environmental stress factor, when applied alone to OFETs, may have no effect on their operation, but when associated with another factor may become active in the formation of defects in the device. Therefore, a complete description of the degradation process in the devices would not be possible by simple degradation experiments, even for devices using well-known and studied organic semiconductors.

An example of the study of the lifetime and defects in OFETs stored under different environmental conditions is shown in PTAA-based transistors (Lau et al. 2013). The examined stress factors are oxygen, water (inert atmosphere, dry air and humid air) and temperature (25–85°C), and the measured degradation parameters over time (up to 5,000 h) are carrier mobility, on/off ratio and threshold voltage. In dry air and inert atmosphere, the charge carrier mobility, the on/off ratio and the threshold voltage of the devices are relatively constant over time. At high temperature (85°C in air) and high humidity (85°C and 85% RH), the carrier mobility strongly decreases and the degradation is attributed to the formation of shallow traps at the dielectric/semiconductor interface. Meanwhile, the on/off ratio decreases in the time range of 0–600 h, which is attributed to oxygen doping of the semiconductor. Then, it increases in the time range of 600–1,200 h and the process is attributed to the trapping of charge carriers at the dielectric–semiconductor interface. The threshold voltage is relatively stable in dry air, inert atmosphere and at high temperature and high humidity, but is strongly degraded at high temperature in air. The degradation is related to a deep hole trapping process at the semiconductor–dielectric interface, which could be introduced by oxygen doping of the semiconductor or of the dielectric. A complementary study of the aged samples by IR spectroscopy indicated that carbonyl groups are developed in the dielectric during aging and are at the origin of defect formation in aged devices.

In another investigation of the effects of oxygen and moisture stress on pentacene-based OFETs using HMDS-treated silicon oxide as a dielectric, exposure of the devices to oxygen or moisture atmosphere for 30 minutes with no applied voltage had no significant effect on the transistor characteristics (Risteska et al. 2016). On the contrary, with an applied bias during the exposure to oxygen or moisture for the same time period, a shift of the threshold voltage is observed, suggesting the formation of active defects in the dielectric/semiconductor interface region. Using a simulation model of the transfer curves, assuming a uniform distribution of defects, under exposure to oxygen and biasing of the device, a formation of acceptor-like defect states occurs at 0.29 eV above the valence band. Exposure of the device under biasing to moisture atmosphere leads to the formation of acceptor-like defect states at 0.34 and 0.80 eV above the valence band.

The device temperature is an important environmental factor of degradation of OFETs. As an example, the role of temperature in device degradation has been investigated in OFETs using DH4T as an organic semiconductor and PMMA as a dielectric (Wrachien et al. 2015). Thermal stress was applied to the device brought to the test temperature T_{stress} (from 35 to 80°C) for 1,000 s with no applied bias. After being brought to the measurement temperature T = 20°C, the transfer characteristic was recorded. A decrease in the threshold voltage V_T variation was observed when the device temperature increased, and at higher temperature setting, ΔV_{TH} is recovered reaching its initial value (Figure 6.10). The variation of V_T with temperature is explained by the trapping of the charge carriers at lower temperatures and their release by thermal emission from trapping centers at higher temperatures. The degradation of the device was different when a bias stress ($V_{GS} = -80V, V_{DS} = 0V$) is applied simultaneously with the thermal stress (constant voltage stress or CVS procedure). The decrease of the variation of V_T in this case is large, and there is no recovery at high temperature, indicating a permanent degradation of the device. It has been suggested that formation of trapping centers in both PMMA and DH4T is responsible for the bias stress degradation, which is accelerated by thermal stress.

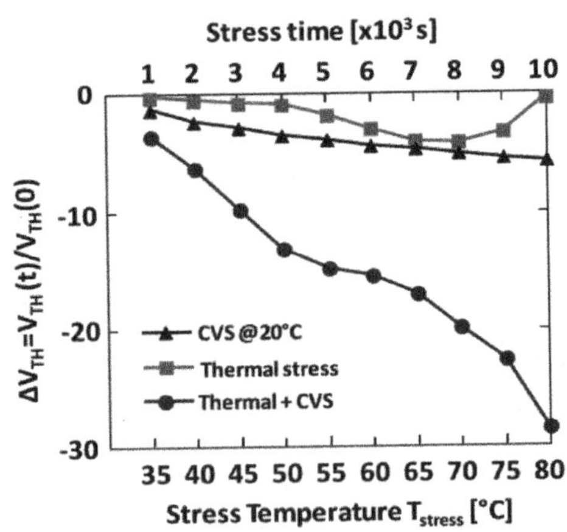

Figure 6.10. *Evolution of the ΔV_{TH} under different stress conditions. (Wrachien et al. 2015, p. 1790). For a color version of this figure, see www.iste.co.uk/nguyen/defects.zip*

Under light illumination, there are two physical processes that can be detected in the OFET responses. The first process is the *photoconductive effect*, which is an increase in the channel conductance, which resulted from the generation of (more mobile) charge carriers following the absorption of photons by the organic semiconductor. Consequently, under irradiation of the device, the drain current intensity I_D increases by photons with energy higher than the band gap energy of the semiconductor. The second process is the *photovoltaic effect*, which is a shift of the threshold voltage V_T and the switch-on voltage V_O, which resulted from the accumulation of the trapped (less mobile) charge carriers in the semiconductor/dielectric interface, producing a hysteresis of the transfer characteristic. Indeed, when the carriers become trapped at the semiconductor/dielectric interface, they can screen the gate field and then induce a shift of the threshold voltage. Consequently, after switching off the irradiation, the drain current intensity I_D should progressively recover its original value when the trapped charge carriers are all released, although this recovery is not always observed because of the slow rate of the trap release. While the first process is fast, the second is slow and its duration can exceed many days.

As an example, the effect of light irradiation of sexithiophene and pentacene-based OFETs was studied using UV light of wavelength in the range 300–400 nm (Noh et al. 2006). Under light irradiation, both types of OFETs show transfer characteristics with photoconductive (increase of the drain current upon light switch-on) and photovoltaic (recovery of the drain current after light switch-off) effects involving electrons and holes with trapping centers. By supposing that the carrier trapping occurs in the channel at the semiconductor/dielectric interface and measuring the maximum shift of V_T, the surface trap density n_{it} can be estimated by (Liguori et al. 2016)

$$\Delta V_T = V_{TL} - V_T = \frac{|e| n_{it}}{C_i} \qquad [6.15]$$

where C_i is the dielectric capacitance. In pentacene-based OFETs using ActivInk D0150 as a dielectric, the surface trap density is determined to be $\sim 9 \times 10^{11} cm^{-2}$. It is also demonstrated in these devices that the combination of bias and light stress lead to either an acceleration of the threshold voltage or a suppression of the photovoltaic effect depending on the polarity of the gate bias.

Other consequences of exposure OFETs to light is the enhancement of some environmental stresses. For example, metal–insulator–semiconductor (MIS) diodes of structure $ITO/Resist/P3HT/Au$ were characterized by capacitance–voltage measurements as a function of ambient and illumination. Under exposure to dry oxygen in the dark, the initial acceptor density of $\sim 5 \times 10^{15} cm^{-3}$ was found to be relatively stable and in the presence of light, a strong increase in the acceptor density was observed and reached $\sim 1.5 \times 10^{16} cm^{-3}$.

It should be noted that under intense light exposure, the degradation of devices is irreversible, and their lifetime is strongly reduced. For instance, unencapsulated P3HT-based OFET of structure $PET/PMMA/P3HT/Au$ shows no detectable degradation under exposure to ambient light for $1{,}000\ h$ (Ficker et al. 2004). When the devices were illuminated by an AM 1.5G solar spectrum light of intensity of $110\ mW.cm^{-2}$ under ambient atmosphere, their lifetime was strongly decreased to $T_{50}{\sim}15\ h$. The degradation is attributed to the alteration of the polymer conjugation length by the reaction of singlet oxygen, which forms carbonyl functional groups.

6.2.2.3.3. Reliability of OFETs

As it can be seen, the investigations of OFET reliability are complicated not only because of the nature and the quality of the numerous organic semiconductors and the dielectric materials used as the active and the insulator layers, but also because of the numerous possibilities to build the devices' architectures, making the comparison of the characterization results and therefore, the identification of degradation processes that are of primary importance for understanding the role of defects in the reliability of OFETs, difficult. The encountered investigation difficulties are also common to other organic devices (OLEDs and OPVs). However, it appears that the complexity of the studied materials and their device structures are less important compared to OFETs. For instance, in the field of OPVs, most of the studies were concentrated to the P3HT:PCBM blends used as absorber materials and more recently, to the perovskite, which is promising for high efficiency and long lifetime. The architecture of these devices is almost similar, having a conventional structure ($ITO/HTL/absorber/ETL/metal$), which makes it possible to study the role of each layer in the formation of defects and their influence on the degradation processes. This is not the case in the OFET field.

As an example, consider the architecture of an OFET using a given organic semiconductor as the active layer and a dielectric as an insulator layer. There are four possible architectures for realizing the device, which are (i) bottom gate, bottom contact (BGBC), (ii) bottom gate, top contact (BGTC), (iii) top gate, bottom contact (TGBC) and (iv) top gate, top contact (BGBC). Of course, each architecture has advantages and disadvantages, and will be adopted according to the chosen purpose (easy deposition, relative protection from environmental stress, etc.). This structure difference will necessitate an adaptation of the characteristic measurement protocol and of the result analysis to obtain a coherent interpretation between studies carried out in different groups. Furthermore, contrarily to other organic devices, there are only a few results on the determination of defects in OFETs using electrical techniques, such as TSC or DLTS. A quantitative characterization of the defect parameters is important to understand the nature of defects, and how they can affect the device performance and lifetime. It is evident that the description of defects with information on their characteristic parameters is much more accurate than a simple statement on the presence of trap states in the material or at the material/contact layer interface, even if this can be useful for the defect investigation. There are several techniques that can be adapted and used to determine the density of states and other trap parameters in transistor structures (energy activation, capture cross-section). For instance, the DLTS technique applied to pentacene diodes of structure $Au/pentacene/SiO_2/Au$ (Yang et al. 2002) allowed us to determine three trap levels of 0.24, 0.31 and 1.08 eV above the valence band edge with concentrations in the range $10^{14} - 10^{15} cm^{-3}$. The TSC technique is also used to characterize trap states in P3HT diodes of structure $Au/Ag/P3HT/SiO_2/Si/Al$ (Rojek et al. 2019), which allows us to determine electron and hole trap state distribution in the organic semiconductor. The hole trap density distribution exhibits a maximum at 100 meV and a shoulder over the range of 200–500 meV, with a density of $2.5 \times 10^{15} cm^{-3}$. The electron trap density distribution shows two maximum peaks centered at 100 and 400 meV with a density of $1.3 \times 10^{15} cm^{-3}$. Other characterization techniques are available for the analysis of defects in OFETs, although not all of them enable access to the description of basic trap parameters, which are of primary importance for the defect study in devices. Among these techniques, the Grünewald method (Grünewald et al. 1980) allows for the determination of the DOS of the organic semiconductor from the transfer characteristic of the transistor. For example, the method was applied to study the effects of temperature on the trap state density in organic small molecule/polymer blend transistors, which shows a broadening distribution when the device temperature increases (Hunter et al. 2016).

From the results of defect investigation in OFETs, it appears that bias stress has the most important effect on the degradation of organic transistors. The physical process is caused by the charge carrier trapping by defects created inside the devices. These defects can originate from any layer of the device or diffuse into the transistor from external environmental media. In particular, the effects of the bias stress are generally accelerated by other environmental stresses (oxygen, water, heat and light) when the device is simultaneously exposed to these stresses during its operation, resulting in a faster degradation of the transistor.

Future Prospects

Investigations of the properties and physical processes occurring in organic semiconductors and devices have usually led researchers to compare organic and inorganic materials over many fields. Some comparison considerations are pertinent and justified; others are less so. It could be seen that there is almost a wide consensus over the primary role of defects in the determination of the performance and the lifetime of both the organic and inorganic materials and devices. Though because of their material structure and their particular device architecture, the formation mechanisms and the energetic distribution of the defects in organic semiconductors and devices differ in many aspects from those in the inorganic counterparts. Consequently, the degradation processes related to defects for these cases may also occur in different ways as commonly observed in inorganic materials and devices. This difference means that the reliability issues of organic components have not yet been solved to a satisfactory degree, despite many successful efforts to bring the new organic electronic devices into the market.

High reliability of devices results in part from the understanding and identification of the degradation mechanisms which occur inside the electronic components, and from the suppression of the responsible degradation sources. Simultaneously, the improvement of the performance and the stability of materials and devices is important to achieve high reliability. Indeed, the reliability of electronic components in the inorganic field has been solidly established, favored by the fact that the selected materials and the chosen architectures of devices have been largely adopted by the manufacturers. Silicon is the main semiconductor for the fabrication of basic electronic components and for devices such as solar cells. According to the National Renewable Energy Laboratory's (NREL) best research efficiency chart in 2022, the efficiency of single crystal silicon-based cells is ~26% and the cell lifetime is estimated to be in the range of 15–20 years. Gallium and indium compound semiconductors are commonly used as emitters in light-emitting devices with a large covering emission of the visible spectrum. A high external quantum efficiency (EQE) of ~18% (Wang et al. 2021) is obtained in InGaN quantum dot-based LEDs and the lifetime of inorganic LEDs is usually in the range of 25,000–50,000 h. (Yazdan Mehr et al. 2020).

For organic devices, the performance and the lifetime of some light-emitting diodes and solar cells are quite comparable to those of the best inorganic counterparts, as shown by the examples cited in the previous sections. The difference, however, exists between the two material fields. Whereas inorganic devices have already been commercialized and have occupied nearly all the markets for transistors, lighting and solar energy, only a few applications of OLEDs, OPVs and OFETs of weak power are available at present. There are several explanations for this apparent contradiction. Firstly, the high performance and long lifetime of the studied organic devices have been obtained for the devices built in laboratory, under strictly well-controlled fabrication and measurement conditions. When these requirements are not met, for instance with a mass production, the device characteristics will be degraded. Secondly, for practical applications, the individual devices should be packaged into modules of a large active area (for solar cells, for example). This integration usually leads to a loss of performance due to the cell interconnection and additionally an increase in degradation by mechanical and electrical stresses. Thirdly, in real working conditions, devices are exposed to environmental stresses (light, heat, humidity) and their characteristics may be strongly altered. Extrapolation of results obtained by accelerated indoor testing may be not sufficient and should be correlated or completed by outdoor measurements, which are still not always performed in characterization investigations. As a consequence, improvements of long-term stability have to be realized for expecting commercially viable organic devices.

We have seen through the investigation examples of OLEDs, OPVs and OFETs that the defects, degradation, performance and the lifetime of the organic devices are closely linked. Many suggestions have been made for the improvement of the reliability of electronic organic devices. These suggestions cover the general as well as the specific aspects of materials, devices, processing and operations that can lead to premature degradation and low reliability of the organic components, and it is not possible to mention them all here. In short, we can summarize the following issues that may impact on the reliability and the future development of organic semiconductors and devices. Four points are to be considered for improving the reliability:

1) Better quality materials of small molecules, conjugated polymers and hybrid materials are needed. Principally, these improved materials should have a low defect density since the defects affect the performance and the stability of organic devices. They should also have a high glass transition temperature T_G to avoid structural instability which may accelerate intrinsic degradation processes. In particular, for flexible devices, a low tensile modulus and a high crack-onset strain are necessary

for withstanding a repeated deformation. Accessorily, depending on the applications, other physical properties need to be improved for enhancing the performance and the lifetime of devices. For instance, a better light absorption would be desired for OPV absorber materials, and a better carrier mobility would be needed for an OFET semiconductor and an OLED emitter. Several currently used organic semiconductors are highly efficient but suffer from premature degradation, while others are more robust against instability but are less performant. Future research is expected to take advantage of both these two types of studied materials. Improvement in the quality of the material will necessarily be accompanied by an improvement in the processing technology to obtain thin films of large area and uniformity.

2) Better characterization of defects in organic semiconductors and devices is necessary to understand their formation, their localization, their size and distribution, and their evolution during the device operation and during their lifetime. There are several techniques for measuring defects in devices and several approaches for determining their parameters. These approaches are useful and efficient as they have allowed us to understand many aspects of the defect–degradation correlation in organic materials and devices. For instance, in some experiments, by exposing the organic devices to a specific stress and by investigating the changes after degradation in their optical, structural, morphological and electrical properties, it is possible to draw a conclusion on the cause of the observed degradation and to relate it to the changes in defect states. By using this approach and for some cases, the role of interfaces between layers as well as the effects of the stress sources in the degradation processes has been in part demonstrated to be closely linked to the formation and development of defects. It is expected that this kind of approach will be more systematically used to shed additional light on the degradation processes for a more complete and consistent description of the defect–degradation relationship. It should be emphasized that only a few measurement techniques have been used to determine the defect parameters quantitatively and accurately in materials and devices though these parameters are quite important for estimating the influence of the defects investigated with the view to understanding the degradation processes (Sirringhaus 2009). For this purpose, a number of specific electrical measurements such as TSC and DLTS can and should be used to quantitatively determine the defect parameters of organic devices and should be exploited more often in future investigations to determine the predominant degradation mechanisms during the device lifetime.

3) Better understanding of the degradation mechanisms of organic semiconductors and devices deposited on flexible substrates needs to be accomplished. Flexibility is one of the major assets of organic devices, which allow for their use in rollable or foldable displays, flexible solar cells and printed electronic components (memories, sensors) on various types of surface shapes. For the applications, flexible devices have a similar architecture to that of their rigid counterparts but all the constituent layers including the substrates (Di Giacomo et al.

2016), the electrodes (Kwon et al. 2017) and the encapsulation (Sutherland et al. 2021) are flexible. The degradation processes in these devices are the same as those occurring in rigid devices, but the use of specific materials for assuring the flexibility may introduce additional stress sources and weaken their overall stability. In addition, mechanical stress in organic materials may also induce a morphological change of the layer interfaces, delamination of the active layer, and metal electrodes, punctures, and cracks in all layers. Generally, repeated mechanical deformation (fatigue stress), even of weak intensity, produces cracks in a metal film deposited on a polymer substrate and increases its electrical resistance, altering the performance of the device. Cracks are believed to be created around defect sites, which are associated with the morphology of the layer and progressively propagate and develop along the layer interface (Harris et al. 2016). In complex device architectures with barrier, dielectric, active, transport and electrode layers deposited on a flexible substrate, degradation occurs by the formation of defects in mechanical cohesive cracks within layers and debonding of interfaces (Brand et al. 2012). These induced changes would enhance the ingress of oxygen and moistures, and as a result, increase the degradation rate of the flexible device. From the results of recent investigations in flexible electronics, the progress in material and structure engineering has significantly improved the performance and lifetime of flexible devices (Root et al. 2017). However, the evolution of the structure and the defect generation of the organic materials in devices under strain are still poorly understood and need further investigations in order to find solutions for these effects and to assure the long-term reliability against mechanical stress.

4) Better definitions of standards for studying the reliability of organic electronic components, that is, their performance and lifetime in relation to degradation and defects, especially in OLEDs and OFETs, are needed. Comparisons between different results of OPV investigations have been greatly facilitated since the agreement on the ISOS testing protocols, which precisely defined the measurement conditions for testing the stability and the lifetime of organic solar cells. The main advantage of these protocols lies in uncovering newly emerging or still to be discovered degradation mechanisms, which then will allow for improving the stability of devices. In the OLED and OFET fields, such a protocol agreement has not been established yet and there are real difficulties to compare the results reported in the literature from different groups, and therefore, difficulties to accurately identify the factors or mechanisms which may affect the degradation of devices. It is noted that at present, there are no established protocols for stability and lifetime measurements of perovskite-based solar cells. Taking, for example, the lifetime measurements of OLEDs, which should be performed by recording the luminance over time and measuring the elapsed time between the initial instant T_0 and the instant T_{50} (or similar values). To enable comparison between the devices, the initial luminance should be set to an identical value for the operational conditions (for instance 100 or 1,000 $cd.m^2$ depending on the use of OLEDs, for display or for

lighting). If this required condition, among others such as environment or temperature, is not fulfilled, the lifetime of similar devices would show large discrepancies, and its evaluation becomes complicated. Several suggestions for measuring the device performance and determining their stability and lifetime have been proposed (Anaya et al. 2019). However, the testing standard should be better described and a global agreement of the concerned research community on the standardization protocols should be established to guide the research on the organic devices to obtain a better reliability level, even for OLEDs, which have already reached the commercialization stage.

We live in a world of perpetual change. The landscape in organic electronics has greatly evolved since the realization of the first thin film OLEDs by Tang and Van Slyke (1987). From the research efforts in laboratories, considerable advances in knowledge of the physical process in organic semiconductors and devices in relation to their structure and architecture have allowed for the realization and then the commercialization of several organic electronic devices for everyday use in displays, lighting, solar cells and sensors. Although the reliability of the commercialized devices is acceptable, it should be improved to reach the level of their inorganic counterparts to expect a large development and expansion in the future. Defect states seem to be an important factor for the performance and lifetime of the organic devices, which in turn influence their reliability. With the knowledge acquired from the inorganic semiconductor field, investigations of defects in organic devices have been successfully carried out by applying and adapting known measurement techniques to new materials and structures and by using new methods specially adapted to organic devices. In this way, the knowledge of defects has been deepened with information concerning how and where they are formed inside the organic structure and which device characteristics they will affect and thus, influence the reliability. The role of the material compositions, processing techniques and environmental stresses in the defect formation, in correlation with the performance and lifetime measurements, has been investigated and analyzed. Some remarkable findings are summarized as follows. It appears that the use of an efficient barrier material for encapsulation is mandatory to prevent organic devices from degradation with alteration of performance and lifetime. The encapsulation of thin films slows down the diffusion rate of water and moistures from the external medium into the active materials, where they act as defect sites changing the charge carrier transport and altering the operational characteristics of the device. In addition, compared with crystalline semiconductor devices, the defect sources are more various in organic materials and the defect sites are not discrete but distributed within the band gap of the semiconductors. Generally, the degradation of the materials produces changes in the distribution function, modifying its intensity, shape and energy. Besides, the role of mechanical stresses on the formation of defects in flexible devices has been addressed. Despite their impact on the working condition of flexible components, these defects have not been yet well understood and as they potentially affect the performance and the lifetime of devices, they should be further investigated in the future.

There is no doubt that organic devices will play an important role in several sectors of electronics in the near future, with the progress made for better understanding of the structure–property relation in these materials and devices. In this direction, this book is written as an introduction to the defects in organic semiconductors, giving an overview of recent progresses in research on those materials and does not pretend to provide a complete description of defect processes in organic electronics since the field is vast and the investigations are often complicated. We hope that it will stimulate more research works in this important and exciting field. Like anything else, there will be room for improvement in the understanding of defects in organic materials and devices since we all know that nothing is perfect.

References

Aberle, A.G. (2000). Surface passivation of crystalline silicon solar cells: A review. *Progress in Photovoltaics: Research and Applications*, 8, 473–487.

Adachi, C. (2014). Third generation of organic electroluminescence materials. *Japanese Journal of Applied Physics*, 53, 060101.

Adachi, C., Baldo, M.A., Thompson, M.E., Forrest, S.R. (2001). Nearly 100% internal phosphorescence efficiency in an organic light-emitting device. *Journal of Applied Physics*, 90, 5048–5051.

Ahmad, S. (2014). Organic semiconductors for device applications: Current trends and future prospects. *Journal of Polymer Engineering*, 34, 279–338.

Akin, S., Arora, N., Zakeeruddin, S.M., Grätzel, M., Friend, R.H., Ibrahim Dar, M. (2020). New strategies for defect passivation in high-efficiency perovskite solar cells. *Advanced Energy Materials*, 10, 1903090.

Almeida, A.J., Sahu, A., Riedinger, A., Norris, D.J., Brandt, M.S., Stutzmann, M., Pereira, R.N. (2016). Charge trapping defects in CdSe nanocrystal quantum dots. *Journal of Physical Chemistry C*, 120, 13763–13770.

Anaya, M., Rand, B.P., Holmes, R.J., Credgington, D., Bolink, H.J., Friend, R.H., Wang, J., Greenham, N.C., Stranks, S.D. (2019). Best practices for measuring emerging light-emitting diode technologies. *Nature Photonics*, 13, 818–821.

Anderson, P.W. (1958). Absence of diffusion in certain random lattices. *Physical Review*, 109, 1492–1505.

Angel, F.A., Wallace, J.U., Tang, C.W. (2017). Effect of lithium and silver diffusion in single-stack and tandem OLED devices. *Organic Electronics*, 42, 102–106.

Argawal, S.C. (1974). Analysis of the thermally stimulated capacitor-discharge method for characterizing localized states in amorphous semiconductors. *Physical Review B*, 10, 4340–4350.

Aristidou, N., Sanchez-Molina, I., Chotchuangchutchaval, T., Brown, M., Martinez, L., Rath, T., Haque, S.A. (2015). The role of oxygen in the degradation of methylammonium lead trihalide perovskite photoactive layers. *Angewandte Chemistry*, 127, 8326–8330.

Arkhipov, V.I., Heremans, P., Emelianova, E.V., Adriaenssens, G.J. (2001). Space charge-limited currents in materials with Gaussian energy distributions of localzed states. *Applied Physics Letters*, 79, 4154–4156.

Arkhipov, V.I., Heremans, P., Emelianova, E., Bässler, H. (2005). Effect of doping on the density-of-states distribution and carrier hopping in disordered organic semiconductors. *Physical Review B*, 71, 045214.

Arora, B.M., Chakravarty, S., Subramanian, S., Polyakov, V.I., Ermakov, M.G., Ermakova, O.N. Perov, P.I. (1993). Deep-level transient charge spectroscopy of Sn donors in AlxGa1-x As. *Journal of Applied Physics*, 73, 1802–1806.

Aziz, A. and Narasimhan, K.L. (2002). Subband gap optical absorption and defects in Tris (8 hydroxy quinolato) aluminium. *Synthetic Metals*, 131, 71–77.

Aziz, H., Popovic, Z.D., Hu, N.X., Hor, A.M., Xu, G. (1999). Degradation mechanism of small moleculebased organic light-emitting devices. *Science*, 283, 1900–1902.

Baena, J.P.C., Steier, L., Tress, W., Saliba, M., Neutzner, S., Matsui, T., Giordano, F., Jacobsson, T.J., Kandada, A.R.S., Zakeeruddin, S.M. et al. (2015). Highly efficient planar perovskite solar cells through band alignment engineering. *Energy & Environmental Science*, 8 2928–2934.

Baldo, M.A., O'Brien, D.F., You, Y., Shoustikov, A., Sibley, S., Thompson, M.E., Forrest, S.R. (1998). Highly efficient phosphorescent emission from organic electroluminescent devices. *Nature*, 395, 151–154.

Barancok, D., Kluvanek, P., Thurzo, I., Vajda, J. (1999). How to distinguish between discrete and distributed charge relaxations in charge transient spectroscopy. *Physica Status Solidi (A)*, 172, 519–528.

Baranovskii, S.D. (2014). Theoretical description of charge transport in disordered organic semiconductors. *Physica Status Solidi (B)*, 251, 487–525.

Bässler, H. (1993). Charge transport in disordered organic photoconductors. A Monte Carlo simulation study. *Physica Status Solidi (B)*, 175, 15–56.

Bässler, H., Borsenberger, P.M., Perry, R.J. (1994). Charge transport in poly(methylphenylsialne): The case of superimposed disorder and polaron effects. *Journal of Polymer Science: Part B: Polymer Physics*, 32, 1677–1685.

Battaglia, C., Cuevas, A., De Wolf, S. (2016). High-efficiency crystalline silicon solar cells: Status and perspectives. *Energy & Environment Science*, 9, 1552–1576.

Baumann, A., Väth, S., Rieder, P., Heiber, M.C., Tvingstedt, K., Dyakonov, V. (2015). Identification of trap states in perovskite solar cells. *Journal of Physical Chemistry Letters*, 6, 2350–2354.

Bharathan, J.M. and Yang, Y. (1998). Polymer/metal interfaces and the performance of polymer light-emitting diodes. *Journal of Applied Physics*, 84, 3207–3211.

Bisquert, J. (2008). Beyond the quasistatic approximation: Impedance and capacitance of an exponential distribution of traps. *Physical Review B*, 77, 235203.

Blaksley, J.C. and Neher, D. (2011). Relationship between energetic disorder and open-circuit voltage in bulk heterojunction organic solar cells. *Physical Review*, B84, 075210.

Blood, P. and Norton, J.W. (1978). The electrical characterisation of semiconductors. *Reports on Progress in Physics*, 41, 11–257.

Blood, P. and Orton, J.W. (1992). *The Electrical Characterization of Semiconductors: Majority Carriers and Electron States*. Academic Press, London.

Boix, P.P., Garcia-Belmonte, G., Muñecas, U., Neophytou, M., Waldauf, C., Pacios, R. (2009). Determination of gap defect states in organic bulk heterojunction solar cells from capacitance measurements. *Applied Physics Letters*, 95, 233302.

Bonoldi, L., Carati, C., Montanari, L., Po', R. (2014). Effects of aging and annealing on the density of trap states in organic photovoltaic materials. *Journal of Physical Chemistry C*, 118, 7751–7758.

Booth, A.H. (1954). Calculation of electron trap depths from thermoluminescence maxima. *Canadian Journal of Chemistry*, 32, 214–215.

Borsuk, J.A. and Swanson, R.M. (1980). Current transient spectroscopy: A high-sensitivity DLTS system. *IEEE Transactions on Electron Devices*, 27, 2217–2225.

Brabec, C.J. (2004). Organic photovoltaics: Technology and market. *Solar Energy Materials and Solar Cells*, 83(2–3), 273–292.

Brabec, C.J., Cravino, A., Meissner, D., Sariciftci, N.S., Fromherz, T., Rispens, M.T., Sanchez, L., Hummelen, J.C. (2011). Origin of the open circuit voltage of plastic solar cells. *Advanced Functional Materials*, 11, 374–380.

Brand, V., Bruner, C., Dauskardt, R.H. (2012). Cohesion and device reliability in organic bulk heterojunction photovoltaic cells. *Solar Energy Materials & Solar Cells*, 99, 182–189.

Brenes, R., Guo, D., Osherov, A., Noel, N.K., Eames, C., Hutter, E.M., Pathak, S.K., Niroui, F., Friend, R.H., Islam, M.S. et al. (2017). Metal halide perovskite polycrystalline films exhibiting properties of single crystals. *Joule*, 1, 155–167.

Broniatowski, A., Blosse, A., Srivastava, P.C., Bourgoin, J.C. (1983). Transient capacitance measurements on resistive samples. *Journal of Applied Physics*, 54, 2907–2910.

Bube, R.H. (1960). *Photoconductivity of Solids*. John Wiley & Sons, New York and London.

Bucci, C., Fieschi, R., Guidi, G. (1966). Ionic thermocurrents in dielectrics. *Physical Review*, 148, 816–823.

Burtone, L., Ray, D., Leo, K., Riede, M. (2012). Impedance model of trap states for characterization of organic semiconductor device. *Journal of Applied Physics*, 111, 064503.

Cacovich, S., Ciná, L., Matteocci, F., Divitini, G., Midgley, P.A., Di Carlob, A., Ducatia, C. (2017). Gold and iodine diffusion in large area perovskitesolar cells under illumination. *Nanoscale*, 9, 4700–4706.

Campbell, I.H., Smith, D.L., Ferraris, P. (1995). Electrical impedance measurements of polymer light-emitting diodes. *Applied Physics Letters*, 66, 3030–3032.

Campbell, A.J., Bradley, D.D.C., Lidzey, D.G. (1997). Space-charge limited conduction with traps in poly (phenylene vinylene) light emitting diodes. *Journal of Applied Physics*, 82, 6326–6342.

Campbell, A.J., Bradley, D.D.C., Werner, E., Brütting, W. (2000a). Transient capacitance measurements of the transport and trap states distributions in a conjugated polymer. *Organic Electronics*, 1, 21–26.

Campbell, A.J., Bradley, D.D.C., Werner, E., Brütting, W. (2000b). Deep level transient spectroscopy (DLTS) of a poly(*p*-phenylene vinylene) Schottky diode. *Synthetic Metals*, 111–112, 273–276.

Carr, J.A. and Chaudhary, S. (2013). On the identification of deeper defect levels in organic photovoltaic devices. *Journal of Applied Physics*, 114, 064509.

Chantre, A., Vincent, G., Bois, D. (1981). Deep-level optical spectroscopy in GaAs. *Physical Review B*, 23, 5335–5359.

Chatham, H. (1996). Oxygen diffusion barrier properties of transparent oxide coatings on polymeric substrates. *Surface and Coatings Technology*, 78, 1–9.

Chen, R. and Kirsh, Y. (1981). *Analysis of Thermally Stimulated Processes*. Pergamon Press, Oxford and New York.

Chen, Q., Zhou, H., Fang, Y., Stieg, A.Z., Song, T.B., Wang, H.H., Xu, X., Liu, Y., Lu, S., You, J. et al. (2015). The optoelectronic role of chlorine in CH3NH3PbI3(Cl)-based perovskite solar cells. *Nature Communications*, 6, 7269.

Cheng, Y.J., Yang, S.H., Hsu, C.H. (2009). Synthesis of conjugated polymers for organic solar cell applications. *Chemical Reviews*, 109, 5868–5923.

Cheng, P., Li, Y., Zhan, X. (2013). A DMF-assisted solution process boosts the efficiency in P3HT:PCBM solar cells up to 5.31%. *Nanotechnology*, 24, 484008.

Cheng, P., Li, G., Zhan, X., Yang, Y. (2018). Next-generation organic photovoltaics based on non-fullerene acceptors. *Nature Photonics*, 12, 131–142.

Choi, H.H., Lee, W.H., Cho, K. (2012). Bias-stress-induced charge trapping at polymer chain ends of polymer gate-dielectrics in organic transistors. *Advanced Functional Materials*, 22, 4833–4839.

Cipolloni, S., Mariucci, L., Valletta, A., Simeone, D., De Angelis, F., Fortunato, G. (2007). Aging effects and electrical stability in pentacene thin film transistors. *Thin Solid Films*, 515, 7546–7550.

Clarke, T.M., Peet, J., Nattestad, A., Drolet, N., Dennler, G., Lungenschmied, C., Leclerc, M., Mozer, A.J. (2012). Charge carrier mobility, bimolecular recombination and trapping in polycarbazole copolymer:fullerene (PCDTBT:PCBM) bulk heterojunction solar cells. *Organic Electronics*, 13, 2639–2646.

Clarke, A.J., Luke, J., Meitzner, R., Wu, J., Wang, Y., Lee, H.K.H., Speller, E.M., Bristow, H., Cha, H., Newman, M.J. et al. (2021). Non-fullerene acceptor photostability and its impact on organic solar cell lifetime. *Cell Reports Physical Science*, 2, 100498.

Cohen, M.H., Fritzsche, H., Ovshinsky, S.R. (1969). Simple band model for amorphous semiconducting alloys. *Physical Review Letters*, 22, 1065–1068.

Craciun, N.I., Zhang, Y., Palmaerts, A., Nicolai, H.T., Kuik, M., Kist, R.J.P., Wetzelaer, G.A.H., Wildeman, J., Vandenbergh, J., Lutsen, L. et al. (2010). Hysteresis-free electron currents in poly (p-phenylene vinylene) derivatives. *Journal of Applied Physics*, 107, 124504.

Crispin, X., Geskin, V., Crispin, A., Cornil, J., Lazzaroni, R., Salaneck, W.R., Bredas, J.L. (2002). Characterization of the interface dipole at organic/metal interfaces. *Journal of the American Chemical Society*, 124, 8131–8141.

Cui, Y., Yao, H., Zhang, J., Xian, K., Zhang, T., Hong, L., Wang, Y., Xu, Y., Ma, K., An, C. et al. (2020). Single-junction organic photovoltaic cells with approaching 18% efficiency. *Advanced Materials*, 32, 1908205.

Cui, Y., Xu, Y., Yao, H., Bi, P., Hong, L., Zhang, J., Zu, Y., Zhang, T., Qin, J., Ren, J. et al. (2021). Single-junction organic photovoltaic cell with 19% efficiency. *Advanced Materials*, 33, 2102420.

Dacuna, J. and Salleo, A. (2011). Modeling space charge-limited currents in organic semiconductors: Extracting trap density and mobility. *Physical Review B*, 84, 195209.

Davis, E.A. (1985). Defects and associated electronic states in amorphous semiconductors. *Journal of Non-Crystalline Solids*, 71, 113–123.

De Wolf, S., Holovsky, J., Moon, S.-J., Löper, P., Niesen, B., Ledinsky, M., Haug, F.-J., Yum, J.-H., Ballif, C. (2014). Organometallic halide perovskites: Sharp optical absorption edge and its relation to photovoltaic performance. *The Journal of Physical Chemistry Letters*, 5, 1035–1039.

DeLongchamp, D.M., Kline, R.J., Fischer, D.A., Richter, L.J., Toney, M.F. (2011). Molecular characterization of organic electronic films. *Advanced Materials*, 23, 319–337.

Deng, Y., Keum, C., Hillebrandt, S., Murawski, C., Gather, M.C. (2021). Improving the thermal stability of top-emitting organic light-emitting diodes by modification of the anode interface. *Advanced Optical Materials*, 9, 2001642.

deQuillettes, D.W., Zhang, W., Burlakov, V.M., Graham, D.J., Leijtens, T., Osherov, A., Bulovic, V., Snaith, H.J., Stranks, S.D. (2016). Photo-induced halide redistribution in organic–inorganic perovskite films. *Nature Communications*, 7, 11683.

Di Giacomo, F., Fakharuddin, A., Josec, R., Brown, T.M. (2016). Progress, challenges and perspectives in flexible perovskite solar cells. *Energy & Environmental Science*, 9, 3007–3035.

Djurisic, A.B., Ng, A.M.C., Chen, X.Y. (2010). ZnO nanostructures for optoelectronics: Materials properties and device applications. *Progress in Quantum Electronics*, 34, 191–259.

Dobaczewski, L., Kaczor, P., Hawkins, I.D., Peaker, A.R. (1994). Laplace transform deep-level transient spectroscopic studies of defects in semiconductors. *Journal of Applied Physics*, 76, 194–198.

Dobaczewski, L., Peaker, A.R., Bonde Nielsen, K. (2004). Laplace-transform deep level spectroscopy: The technique and its applications to the study of point defects in semiconductors. *Journal of Applied Physics*, 96, 4689–4728.

Dodabalapur, A., Katz, H.E., Torsi, L., Haddon, R.C. (1996). Organic field-effect bipolar transistors. *Applied Physics Letters*, 58, 1108–1110.

Domanski, K., Correa-Baena, J.P., Mine, N., Nazeeruddin, M.K., Abate, A., Saliba, M., Tress, W., Hagfeldt, A., Gratzel, M. (2016). Not all that glitters is gold: Metal-migration-induced degradation in perovskite solar cells. *ACS Nano*, 10, 6306–6314.

Driver, M.C. and Wright, G.T. (1963). Thermal release of trapped space charge in solids. *Proceedings of the Physical Society*, 81, 141–147.

Droseros, N., Longo, G., Brauer, J.C., Sessolo, M., Bolink, H.J., Banerji. N. (2018). Origin of the enhanced photoluminescence quantum yield in MAPbBr3 perovskite with reduced crystal size. *ACS Energy Letters*, 3, 1458−1466.

Duan, H.S., Zhou, H., Chen, Q., Sun, P., Luo, S., Song, T.B., Bob, B., Yang, Y. (2015). The identification and characterization of defect states in hybrid organic–inorganic perovskite photovoltaics. *Physical Chemistry Chemical Physics*, 17, 112–116.

Duijnstee, E.A., Ball, J.M., Le Corre, V.M., Koster, J.A., Snaith, H.J., Lim, J. (2020). Toward understanding space-charge limited current measurements on metal halide perovskites. *ACS Energy Letters*, 5, 376–384.

Dupuis, A., Chung, P.W.W., Rivaton, A., Gardette, J.L. (2012). Influence of the microstructure on the photooxidative degradation of poly(3-hexylthiophene). *Polymer Degradation and Stability*, 97, 366–374.

Dyakonov, V., Godovsky, D., Meyer, J., Parisi, J., Brabec, C.J., Sariciftci, N.S., Hummelen, J.C. (2001). Electrical admittance studies of polymer photovoltaic cells. *Synthetic Metals*, 124, 103–105.

Efros, A.L. and Shklovskii, B.I. (1975). Coulomb gap and low temperature conductivity of disordered systems. *Journal of Physics C: Solid State Physics*, 8, L49–L51.

Eisenbarth, T., Unold, T., Caballero, R., Kaufmann, C.A., Schock, H.W. (2010). Interpretation of admittance, capacitance-voltage, and current-voltage signatures in thin film solar cells. *Journal of Applied Physics*, 107, 034509.

Ettaik, H., Nguyen, T.P., Lefrant, S. (1993). TSC measurements in Metal/PPP/Metal structures. *Synthetic Metals*, 55–57, 4204–4209.

Faiman, D. (2008). Assessing the outdoor operating temperature of photovoltaic modules. *Progress in Photovoltaics*, 16, 307–315.

Fang, H.H., Adjokatse, S., Wei, H., Yang, J., Blake, G.R., Huang, J., Even, J., Loi, M.A. (2016). Ultrahigh sensitivity of methylammonium lead tribromide perovskite single crystals to environmental gases. *Science Advances*, 2, e1600534.

Farmer, J.W., Lamp, C.D., Meese, J.M. (1981). Charge transient spectroscopy. *Applied Physics Letters*, 41, 1063–1065.

Féry, C., Racine, B., Vaufrey, D., Doyeux, H., Cinà, S. (2005). Physical mechanism responsible for the stretched exponential decay behavior of aging organic light-emitting diodes. *Applied Physics Letters*, 87, 213502.

Ficker, J., von Seggern, H., Rost, H., Fix, W., Clemens, W., McCulloch, I. (2004). Influence of intensive light exposure on polymer field-effect transistors. *Applied Physics Letters*, 85, 1377–1379.

Fisher, J., Tress, W., Kleeman, H., Widmer, J., Leo, K. (2014). Exploiting diffusion currents at Ohmic contacts for traps characterization in organic semiconductors. *Organic Electronics*, 15, 2428–2432.

Fleissner, A., Stegmaier, K., Melzer, C., von Seggern, H., Schwalm, T., Rehahn, M. (2009). Residual halide groups in gilch-polymerized poly(p-phenylene-vinylene) and their impact on performance and lifetime of organic light-emitting diodes. *Chemistry of Materials*, 21, 4288–4298.

Forsythe, E.W., Morton, D.C., Tang, C.W., Gao, Y. (1998). Trap states of tris-8-(hydroxyquinoline) aluminum and naphthylsubstituted benzidine derivative using thermally stimulated luminescence. *Applied Physics Letters*, 73, 1457.

Fu, L., Li, H., Wang, L., Yin, R., Li, B., Yin, L. (2020). Defect passivation strategies in perovskites for an enhanced photovoltaic performance. *Energy & Environmental Science*, 13, 4017–4056.

Fukuda, M. (1991). *Reliability and Degradation of Semiconductor Lasers and Leds*. Artech House, Boston.

Garlick, G.F.J. and Gibson, A.F. (1948). The electron trap mechanism of luminescence in sulphide and silicate phosphors. *Proceedings of the Physical Society*, 60, 574–590.

Gaudin, O., Jackman, R.B., Nguyen, T.P., Le Rendu, P. (2001). Determination of traps in poly (p-phenylene vinylene) light emitting diodes by charge-based deep level transient spectroscopy. *Journal of Applied Physics*, 90, 4196–4204.

Giebink, N.C., D'Andrade, B.W., Weaver, M.S., Mackenzie, P.B., Brown, J.J., Thompson, M.E., Forrest, S.R. (2008). Intrinsic luminance loss in phosphorescent small-molecule organic light emitting devices due to bimolecular annihilation reactions. *Journal of Applied Physics*, 103, 044509.

Gobrecht, H. and Hofmann, D. (1966). Spectroscopy of traps by fractional glow technique. *Journal of Physics and Chemistry of Solids*, 27, 509–522.

Gonzalez-Valls, I. and Lira-Cantu, M. (2009). Vertically aligned nanostructures of ZnO for excitonic solar cells: A review. *Energy & Environmental Science*, 2, 19–34.

Goris, L., Haenen, K., Nesladek, M., Wagner, P., Vanderzande, D., de Schepper, L., D'haen, J., Lutsen, L., Manca, J.V. (2005). Absorption phenomena in organic thin films for solar cell applications investigated by photothermal deflection spectroscopy. *Journal of Materials Science*, 40, 1413–1418.

Goto, G., Yanagisawa, S., Wada, O., Takanashi, H. (1973). Determination of deep-level energy and density profiles in inhomogeneous semiconductors. *Applied Physics Letters*, 23, 150–152.

Götzen, J., Käfer, D., Wöll, C., Witte, G. (2010). Growth and structure of pentacene films on graphite: Weak adhesion as a key for epitaxial film growth. *Physical Review B*, 81, 085440.

Grancini, G., Roldan-Carmona, C., Zimmermann, I., Mosconi, E., Lee, X., Martineau, D., Narbey, S., Oswald, F., De Angelis, F., Graetzel, M. et al. (2017). One-year stable perovskite solar cells by 2D/3D interface engineering. *Nature Communications*, 8, 15684.

Grünewald, M., Thomas, P., Würtz, D. (1980). A simple scheme for evaluating field effect data. *Physica Status Solidi*, 100, K139–K143.

Guha, S., Rice, J.D., Yau, Y.T., Martin, C.M., Chandrasekhar, M., Chandrasekhar, H.R., Guentner, R., de Freitas, P.S., Scherf, U. (2003). Temperature dependent photoluminescence of organic semiconductors with varying backbone conformation. *Physical Review B*, 67, 125204.

Habisreutinger, S.N., Leijtens, T., Eperon, G.E., Stranks, S.D., Nicholas, R.J., Snaith, H.J. (2014). Carbon nanotube/polymer composites as a highly stable hole collection layer in perovskite solar cells. *Nano Letters*, 14, 5561–5568.

Hall, R.N. (1982). Electron-hole recombination in germanium. *Physical Review*, 87, 387.

Han, S.H., Kim, J.H., Jang, J., Cho, S.M., Lee, M.H.S.H., Choo, D.J. (2006). Lifetime of organic thin-film transistors with organic passivation layers. *Applied Physics Letters*, 88, 071106.

Han, Y., Meyer, S., Dkhissi, Y., Weber, K., Pringle, J.M., Bach, U., Spiccia, L., Cheng, Y.B. (2015). Degradation observations of encapsulated planar CH3NH3PbI3 perovskite solar cells at high temperatures and humidity. *Journal of Materials Chemistry A*, 3, 8139–9147.

Hanfland, R., Fischer, M.A., Brutting, W., Wurfel, U., MacKenzie, R.C.I. (2013). The physical meaning of charge extraction by linearly increasing voltage transients from organic solar cells. *Applied Physics Letters*, 103, 063904.

Harris, K.D., Elias, A.L., Chung, H.J. (2016). Flexible electronics under strain: A review of mechanical characterization and durability enhancement strategies. *Journal of Materials Science*, 51, 2771–2805.

Hearing, R.R. and Adams, E.N. (1960). Theory and application of thermally stimulated currents in photoconductors. *Physical Review*, 117, 451–454.

Heath, J.T., Cohen, J.D., Shafarman, W.N. (2004). Bulk and metastable defects in CuIn1xGaxSe2 thin films using drive-level capacitance profiling. *Journal of Applied Physics*, 95, 1000–1010.

Heil, H., Steiger, J., Karg, S., Gastel, M., Ortner, H., von Seggern, H., Stößel, M. (2001). Mechanisms of injection enhancement in organic light emitting diodes through an Al/LiF electrode. *Journal of Applied Physics*, 89, 420–424.

Heo, S., Seo, G., Lee, Y., Lee, D., Seol, M., Lee, J., Park, J.B., Kim, K., Yun, D.J., Kim, Y.S. et al. (2017). Deep level trapped defect analysis in $CH_3NH_3PbI_3$ perovskite solar cells by deep level transient spectroscopy. *Energy & Environmental Science*, 10, 1128–1133.

Hepp, A., Heil, H., Weise, W., Ahles, M., Schmechel, R., von Seggern, H. (2003a). Light-emitting field-effect transistor based on a tetracene thin film. *Physical Review Letters*, 91, 157406.

Hepp, A., von Malm, N., Schmechel, R., von Seggern, H. (2003b). Effects of process parameters on trap distributions in organic semiconductors. *Synthetic Metals*, 138, 201–207.

Hermenau, M., Riede, M., Leo, K., Gevorgyan, S.A., Krebs, F.C., Norrman, K. (2011). Water and oxygen induced degradation of small molecule organic solar cells. *Solar Energy Materials & Solar Cells*, 95, 1268–1277.

Heumueller, T., Burke, T.M., Mateker, W.R., Sachs-Quintana, I.T., Vandewal, K., Brabec, C.J., McGehee, M.D. (2015). Disorder-induced open-circuit voltage losses in organic solar cells during photoinduced burn-in. *Advanced Energy Materials*, 5, 1500111.

Hindeleh, A.M. and Hoseman, R. (1988). Paracrystals representing the physical state of matter. *Journal of Physics C: Solid State Physics*, 21, 4155–4170.

Ho, C.H.Y., Cheung, S.H., Li, H.W., Chiu, K.L., Cheng, Y., Yin, H., Chan, M.H., So, F., Tsang, S.W., So, S.K. (2017). Using ultralow dosages of electron acceptor to reveal the early-stage donor–acceptor electronic interactions in bulk heterojunction blends. *Advanced Energy Materials*, 7(12), 1602360.

Holliday, S., Ashraf, R.S., Wadsworth, A., Baran, D., Yousaf, S.A., Nielsen, C.B., Tan, C.H., Dimitrov, S.D., Shang, Z., Gasparini, N. et al. (2016). High-efficiency and air-stable P3HT-based polymer solar cells with a new non-fullerene acceptor. *Nature Communications*, 7, 11585.

Horowitz, G. (2015). Validity of the concept of band edge in organic semiconductors. *Journal of Applied Physics*, 118, 115502.

Hsieh, H.C., Hsiow, C.Y., Lin, K.F., Shih, Y.C., Wang, L.Y., Renaud, C., Nguyen, T.P. (2018). Analysis of defects and traps in N-I-P layered-structure of perovskite solar cells by charge-based deep level transient spectroscopy (Q-DLTS). *Journal of Physical Chemistry C*, 122, 17601–17611.

Hu, W., Manabe, K., Furukawa, T., Matsumura, M. (2002). Lowering of operational voltage of organic electroluminescent devices by coating indium-tin-oxide electrodes with a thin CuOx layer. *Applied Physics Letters*, 80, 2640–2641.

Hunter, S., Mottram, A.D., Anthopoulos, T.D. (2016). Temperature and composition-dependent density of states in organic small-molecule/polymer blend transistors. *Journal of Applied Physics*, 120, 025502.

Hwang, W. and Kao, K.C. (1976). Studies of the theory of single and double injections in solids with a Gaussian trap distribution. *Solid-State Electronics*, 19, 1045–1047.

Hwang, J., Wan, A., Kahn, A. (2009). Energetics of metal–organic interfaces: New experiments and assessment of the field. *Materials Science and Engineering R*, 64, 1–31.

Igalson, M. and Zabierowsk, P. (2000). Transient capacitance spectroscopy of defect levels in CIGS devices. *Thin Solid Films*, 361–362, 371–377.

Imperia, P., Casu, M.B., Schrader, S., Falk, B., Jandkec, M., Strohriegl, P. (2001). Thermally stimulated processes in heterocyclic materials suitable for heterolayer organic light emitting diodes. *Synthetic Metals*, 124, 83–85.

Ishii, M., Mori, T., Fujikawa, H., Tokito, S., Taga, Y. (2000). Improvement of organic electroluminescent device performance by in situ plasma treatment of indium–tin-oxide surface. *Journal of Luminescence*, 87–89, 1165–1167.

Jackson, W.B., Amer, N.M., Boccara, A.C., Fournier, D. (1981). Photothermal deflection spectroscopy and detection. *Applied Optics*, 20, 1333–1344.

Jasenek, A., Rau, U., Nadenau, V., Schock, H.W. (2000). Electronic properties of CuGaSe2-based heterojunction solar cells. Part II. Defect spectroscopy. *Journal of Applied Physics*, 87, 594–602.

Jeong, Y.S., Troadec, D., Moliton, A., Ratier, B., Antony, R., Veriot, G. (2002). Dielectric study of Alq3 and transport mechanisms. *Synthetic Metals*, 127, 195–200.

Jeong, J., Kim, M., Seo, J., Lu, H., Ahlawat, P., Mishra, A., Yang, Y., Hope, M.A., Eickemeyer, F.T., Kim, M. et al. (2021). Pseudo-halide anion engineering for α-FAPbI3 perovskite solar cells. *Nature*, 592, 381–385.

Johnson, N.M. and Bartelink, D.J. (1979). Constant-capacitance DLTS measurement of defect-density profiles in semiconductors. *Journal of Applied Physics*, 50, 4828–4833.

Johnson, W.C. and Panousis, P.T. (1971). The influence of Debye length on the C-V measurements of doping profile. *IEEE Transactions on Electron Devices*, ED-18, 965–973.

Jones, G.W., Taylor, D.M., Gomes, H.L. (1997). DLTS investigation of acceptor states in P3MeT Schottky barrier diodes. *Synthetic Metals*, 85, 1341–1342.

Juhasz, P., Vary, M., Stuchlikova, L., Harmatha, L., Jakabovic, J., Weis, M. (2015). Characterization of charge traps in pentacene diodes by electrical methods. *Organic Electronics*, 17, 240–246.

Jurchescu, O.D., Baas, J., Palstra, T.T.M. (2004). Effect of impurities on the mobility of single crystal pentacene. *Applied Physics Letters*, 84, 3061–3063.

Juska, G., Arlauskas, K., Viliūnas, M., Genevicius, K., Österbacka, R., Stubb. H. (2000). Charge transport in π-conjugated polymers from extraction current transients. *Physical Review B*, 62, R16235–R16238.

Kadashchuk, A., Ostapenko, N., Zaika, V., Nespurek, S. (1998). Low-temperature thermoluminescence in poly(methyl-phenylsilylene). *Chemical Physics*, 234, 285–296.

Kamat, P.V. and Kuno, M. (2021). Halide ion migration in perovskite nanocrystals and nanostructures. *Account of Chemical Research*, 54, 520–531.

Kang, J.H., da Silva Filho, D., Bredas, J.L., Zhu, X.Y. (2005). Shallow trap states in pentacene thin films from molecular sliding. *Applied Physics Letters*, 86, 152115.

Kawano, K. and Adachi, C. (2009). Evaluating carrier accumulation in degraded bulk heterojunction organic solar cells by a thermally stimulated current technique. *Advanced Functional Materials*, 19, 3934–3940.

Kawano, K., Pacios, R., Poplavskyy, D., Nelson, J., Durrant, J.R., Bradley, D.D.C. (2006). Degradation of organic solar cells due to air exposure. *Solar Energy Materials & Solar Cells*, 90, 3520–3530.

Kazukaukas, V., Tzeng, H., Chen, S.A. (2002). Trap levels and effect of oxygen in poly[2-methoxy-5-(2′-ethyl-hexyloxy)-1,4-phenylene vinylene] diodes. *Applied Physics Letters*, 80, 2017–2019.

Khelifi, S., Decock, K., Lauwaert, J., Vrielinck, H., Spoltore, D., Piersimoni, F., Manca, J., Belghachi, A., Burgelman, M. (2011). Investigation of defects by admittance spectroscopy measurements in poly (3-hexylthiophene):(6,6)-phenyl C61-butyric acid methyl ester organic solar cells degraded under air exposure. *Journal of Applied Physics*, 110, 094509.

Khenkin, M.V., Katz, E.A., Abate, A., Bardizza, G., Berry, J.J., Brabec, C.J., Brunetti, F., Bulović, V., Burlingame, Q., Di Carlo, A. et al. (2020). Consensus statement for stability assessment and reporting for perovskite photovoltaics based on ISOS procedures. *Nature Energy*, 5, 35–49.

Kim, S., Hsu, C., Zhang, C., Skulason, H., Uckert, F., LeCloux, D., Cao, Y., Parker, I. (2004). Degradation of PLEDs and a way to improve device performances. *Journal of Information Display*, 5, 14–18.

Kim, H.S., Mora-Sero, I., Gonzalez-Pedro, V., Fabregat-Santiago, F., Juarez-Perez, E.J., Park, N.G., Bisquert, J. (2013). Mechanism of carrier accumulation in perovskite thin-absorber solar cells. *Nature Communications*, 4, 2242.

Kim, H.S., Seo, J.Y., Park, N.G. (2016). Impact of selective contacts on long-term stability of CH3NH3PbI3 perovskite solar cells. *Journal of Physical Chemistry*, 120, 27840–27848.

Kirchartz, T. (2013). Influence of diffusion on space charge-limited current measurements in organic semiconductors. *Beilstein Journal of Nanotechnology*, 4, 180–188.

Kirchartz, T., Márquez, J.A., Stolterfoht, M., Unold, T. (2020). Photoluminescence-based characterization of halide perovskites for photovoltaics. *Advanced Energy Materials*, 10, 1904134.

Kirov, K.I. and Radev, K.B. (1981). A simple charge based DLTS technique. *Physica Status Solidi (A)*, 68, 711–716.

Kneisel, J., Siemer, K., Luck, I., Bräunig, D. (2000). Admittance spectroscopy of efficient CuInS2 thin film solar cells. *Journal of Applied Physics*, 88, 5474–5481.

Koden, M., Ohnishi, Y., Nishimura, M., Uchida, H. (2007). Emission and degradation mechanism of PLED. *Proceedings of SPIE*, 6655, 66550C1-10.

Kojima, A., Teshima, K., Miyasaka, T., Shirai. Y. (2006). Novel photoelectrochemical cell with mesoscopic electrodes sensitized by lead-halide compounds (2). *Proceedings of the 210th ECS Meeting*, MA-02, 397.

Kojima, A., Teshima, K., Shirai, Y., Miyasaka, T. (2009). Organometal halide perovskites as visible-light sensitizers for photovoltaic cells. *Journal of American Chemical Society*, 131, 6050–6051.

Kondakov, D.Y., Sandifer, J.R., Tang, C.W., Young, R.H. (2003). Nonradiative recombination centers and electrical aging of organic light-emitting diodes: Direct connection between accumulation of trapped charge and luminance loss. *Journal of Applied Physics*, 93, 1108–1119.

Kondakov, D.Y., Lenhart, W.C., Nichols, W.F. (2007). Operational degradation of organic light-emitting diodes: Mechanism and identification of chemical products. *Journal of Applied Physics*, 101, 024512.

Koster, L.J.A., Mihailetchi, V.D., Xie, H., Blom, P.W.M. (2005). Origin of the light intensity dependence of the short-circuit current of polymer/fullerene solar cells. *Applied Physics Letters*, 87, 203502.

Kuik, M., Vandenbergh, J., Goris, L., Begemann, E.J., Lutsen, L., Vanderzande, D.J.M., Manca, J.V., Blom, P.W.M. (2011). Optical detection of deep electron traps in poly(p-phenylene vinylene) light-emitting diodes. *Applied Physics Letters*, 99, 183305.

Kumar, V., Jain, S.C., Kapoor, A.K., Geens, W., Aernauts, T., Poortmans, J., Mertens, R. (2003). Trap density in conducting organic semiconductors determined from temperature dependence of J-V characteristics. *Journal of Applied Physics*, 94, 1283–1285

Kuniyoshi, S., Naruge, S., Iizuka, M., Nakamura, M., Kudo, K., Tanaka, K. (2003). Thermally stimulated current of pentacene Schottky diode. *Synthetic Metals*, 137, 895–896.

Kwon, J.H., Choi, S., Jeon, Y., Kim, H., Chang, K.S., Choi, K.C. (2017). Functional design of dielectric metal dielectric-based thin-film encapsulation with heat transfer and flexibility for flexible displays. *ACS Applied Materials & Interfaces*, 9, 27062–27072.

Labastide, J.A., Baghgar, M., McKenna, A., Barnes, M.D. (2012). Time- and polarization-resolved photoluminescence decay from isolated polythiophene (P3HT) nanofibers. *Journal of Physical Chemistry*, 116, 23803–23811.

Laird, D.W., Vaidya, S., Li, S., Mathai, M., Woodworth, B., Sheina, E., Williams, S., Hammond, T. (2007). Advances in plexcore active layer technology systems for organic photovoltaics: Roof-top and accelerated lifetime analysis of high performance organic photovoltaic cells. *Proceedings SPIE*, 6656, 66560X.

Lang, D.V. (1974). Deep-level transient spectroscopy: A new method to characterize traps in semiconductors. *Journal of Applied Physics*, 45, 3023–3032.

Lau, T., Lorenz, E., Koyuncu, M. (2013). Lifetime of poly(triaryl amine) based organic field effect transistors under different environmental conditions. *Japanese Journal of Applied Physics*, 52, 041601.

Lee, K.O. and Gan, T.T. (1977). Space charge-limited currents in evaporated films of pentacene. *Physica Status Solidi A*, 43, 565–571.

Lee, S.T., Gao, Z.Q., Hung, L.S. (1999). Metal diffusion from electrodes in organic light-emitting diodes. *Applied Physics Letters*, 75, 1404–1406.

Lee, J., Cohen, J.D., Shafarman, W.N. (2005). The determination of carrier mobilities in CIGS photovoltaic devices using high-frequency admittance measurements. *Thin Solid Films*, 480–481, 336–340.

Lee, C.W., Renaud, C., Le Rendu, P., Nguyen, T.P., Seneclauze, B., Ziessel, R., Kanaan, H., Jolinat, P. (2010). Performance and defects in phosphorescent organic light-emitting diodes. *Solid State Sciences*, 12, 1873–1887.

Lee, K.H., Schwenn, P.E., Smith, A.R.G., Cavaye, H., Shaw, P.E., James, M., Krueger, K.B., Gentle, I.R., Meredith, P., Burn, P.L. (2011). Morphology of all-solution-processed "bilayer" organic solar cells. *Advanced Materials*, 23, 766–770.

Lee, S., Kim, H., Kim, Y. (2021). Progress in organic semiconducting materials with high thermal stability for organic light-emitting devices. *InfoMat*, 3, 61–81.

Lenz, J., del Giudice, F., Geisenhof, F.R., Winterer, F., Weitz, R.T. (2019). Vertical, electrolyte-gated organic transistors show continuous operation in the MA cm^{-2} regime and artificial synaptic behaviour. *Nature Nanotechnology*, 14, 579–585.

Lewandowski, A.C. and McKeever, S.W.S. (1991). Generalized description of thermally stimulated processes without the quasi-equilibrium approximation. *Physical Review B*, 43, 8163–8177.

Li, J.V., Nardes, A.M., Liang, Z., Shaheen, S.E., Gregg, B.A., Levi, D.H. (2011). Simultaneous measurement of carrier density and mobility of organic semiconductors using capacitance techniques. *Organic Electronics*, 12, 1879–1885.

Li, Z., Xiao, C., Yang, Y., Harvey, S.P., Kim, D.H., Christians, J.A., Yang, M., Schulz, P., Nanayakkara, S.U., Jiang, C.-S. et al. (2017). Extrinsic ion migration in perovskite solar cells. *Energy & Environmental Science*, 10, 1234–1242.

Li, Y., Huang, X., Ding, K., Sheriff Jr, H.K.M., Ye, L., Liu, H., Li, C.Z., Ade, H., Forrest, S.R. (2021). Non-fullerene acceptor organic photovoltaics with intrinsic operational lifetimes over 30 years. *Nature Communications*, 12, 5419.

Liguori, R., Sheets, W.C., Facchetti, A., Rubino, A. (2016). Light- and bias-induced effects in pentacene-based thin film phototransistors with a photocurable polymer dielectric. *Organic Electronics*, 28, 147–154.

Lin, Y., Wang, J., Zhang, Z.G., Bai, H., Li, Y., Zhu, D., Zhan, X. (2015). An electron acceptor challenging fullerenes for efficient polymer solar. *Cells Advanced Materials*, 27 1170–1174.

Lin, T., Chatterjee, T., Tsai, W., Lee, W., Wu, M., Jiao, M., Pan, K., Yi, C., Chung, C., Wong, K. et al. (2016). Sky-blue organic light emitting diode with 37% external quantum efficiency using thermally activated delayed fluorescence from spiroacridine-triazine hybrid. *Advanced Materials*, 28, 6976–6983.

List, E.J.W., Guentner, R., Scanducci de Freitas, P., Scher, U. (2002). The effect of keto defect sites on the emission properties of polyfluorene-type materials. *Advanced Materials*, 14, 374–378.

Liu, J., Wang, C.H., Liu, C.H., Li, Q.L., Gao, X., Wang, S.D. (2016). Bias-stress-stable low-voltage organic field-effect transistors with ultrathin polymer dielectric on C nanoparticles. *Advanced Electronic Materials*, 2, 1500349.

Losee, D.L. (1975). Admittance spectroscopy of impurity levels in Schottky barriers. *Journal of Applied Physics*, 46, 2204–2214.

Ma, L. and Yang, Y. (2004). Unique architecture and concept for high-performance organic transistors. *Applied Physics Letters*, 85, 5084–5086.

Macdonald, J.R. (1987). *Impedance Spectroscopy*. John Wiley & Sons, New York.

Macphail, R., Strauss, H., Snyder, R., Ellice, C. (1984). C-H stretching modes and the structure of n-Alkyl chains 2: Long, all-trans chains. *The Journal of Physical Chemistry*, 88, 334–341.

Mandoc, M.M., Kooistra, F.B., Hummelen, J.C., de Boer, B., Blom, P.W.M. (2007a). Effect of traps on the performance of bulk heterojunction organic solar cells. *Applied Physics Letters*, 91, 263505.

Mandoc, M.M., Veurman, W., Koster, L.J.A., de Boer, B., Blom, P.W.M. (2007b). Origin of the reduced fill factor and photocurrent in MDMO-PPV. *Advanced Functional Materials*, 17, 2167–2173.

Mantri, P., Rizvi, S.M.H., Mazhari, B. (2013). Estimation of built-in voltage from the steady-state current-voltage characteristics of organic diodes. *Organic Electronics*, 14, 2034–2038.

Martin, D.C., Wilson, P.M., Liao, J., Jones, M.C.G. (1995). Chain-end defects in extended-chain polymer solids. *MRS Bulletin*, 47–50.

Marumoto, K., Fujimori, T., Ito, M., Mori, T. (2012). Charge formation in pentacene layers during solar-cell fabrication: Direct observation by electron spin resonance. *Advanced Energy Materials*, 2, 591–597.

Matthiew, X. (2003). Photo-induced current transient spectroscopic study of the traps in CdTe. *Solar Energy Materials& Solar Cells*, 76, 225–242.

McKeever, S.W.S. (1981). *Thermoluminescence of Solids*. Cambridge University Press, Cambridge.

McKeever, S.W.S. and Hughes, D.M. (1975). Thermally stimulated currents in dielectrics. *Journal of Physics D: Applied Physics*, 8, 1520–1529.

McKeever, S.W.S. and Sholom, S. (2021). Trap level spectroscopy of disordered materials using thermoluminescence: An application to aluminosilicate glass. *Journal of Luminescence*, 234, 117950.

Meerheim, R., Scholz, S., Olthof, S., Schwartz, G., Reineke, S., Walzer, K., Leo, K. (2008). Influence of charge balance and exciton distribution on efficiency and lifetime of phosphorescent organic light-emitting devices. *Journal of Applied Physics*, 104, 014510.

Michelson, C.E., Gelatos, A.V., Cohen, J.D. (1985). Drive-level capacitance profiling: Its application to determining gap state densities in hydrogenated amorphous silicon films. *Applied Physics Letters*, 47, 412–414.

Miller, A. and Abrahams, E. (1960). Impurity conduction at low concentrations. *Physical Review*, 120, 745–755.

Min, H., Lee, D.Y., Kim, J., Kim, G., Lee, K.S., Kim, J., Paik, M.J., Kim, Y.K., Kim, K.S., Kim, M.G. et al. (2021). Perovskite solar cells with atomically coherent interlayers on SnO2 electrodes. *Nature*, 598, 444–450.

Mingebach, M., Deibel, C., Dyakonov, V. (2011). Built-in potential and validity of the Mott-Schottky analysis in organic bulk heterojunction solar cells. *Physical Review B*, 84, 153201.

Mitonneau, A., Mircea, A., Martin, G.M., Pons, D. (1979). Electron and hole capture cross-section at deep centers in gallium arsenide. *Revue de physique appliquée*, 14, 853–861.

Mizuo, S., Tabatake, S., Naka, S., Okada, H. (2002). Evaluation of bulk trap density of tris-(8-hydroxyquinoline) aluminum. *Journal of Applied Physics*, 92, 1450–1452.

Mladenovic, M. and Vukmirovic, N. (2015). Charge carrier localization and transport in organic semiconductors: Insights from atomistic multiscale simulations. *Advanced Functional Materials*, 25, 1915–1932.

Mooney, P.M. (1983). Photo-deep level transient spectroscopy: A technique to study deep levels in heavily compensated semiconductors. *Journal of Applied Physics*, 54, 208–213.

Mort, J. (1981). Geminate and non-geminate recombination in amorphous semiconductors. *Journal de physique*, 42(C4), 433–441.

Mott, N.F. (1968). Conduction in glasses containing transition metal ions. *Journal of Non-Crystalline Solids*, 1, 1–17.

Mott, N.F. (1980). States in the gap of non-crystalline semiconductors. *Journal of Physics C: Solid State Physics*, 13, 5433–5471.

Mott, N.F. and Davis, E.A. (1979). *Electronic Processes in Non-crystalline Materials*. Oxford University Press, Oxford.

Mott, N.F. and Gurney, R.W. (1940). *Electronic Processes in Ionic Crystals*. Oxford University Press, New York.

Murano, S., Burghart, M., Birnstock, J., Wellmann, P., Vehse, M., Werner, A., Canzler, T., Stübinger, T., He, G., Pfeiffer, M. et al. (2005). Highly efficient white OLEDs for lighting applications. *Proceedings SPIE Organic Light-Emitting Materials and Devices IX*, 5937, 59370H.

Murgatroyd, P.N. (1970). Theory of space-charge-limited current enhanced by Frenkel effect. *Journal of Physics D: Applied Physics*, 3, 151–156.

Nagase, T. and Naito, H. (2000). Localized-state distributions in molecularly doped polymers determined from time-of-flight transient photocurrent. *Journal of Applied Physics*, 88, 252–259.

Naito, H., Ding, J., Okuda, M. (1994). Determination of localized-state distributions in amorphous semiconductors from transient photoconductivity. *Applied Physics Letters*, 64, 1830–1832.

Nan, G., Zhang, X., Abdi-Jalebi, M., Andaji-Garmaroudi, Z., Stranks, S.D., Lu, G., Beljonne, D. (2020). How methylammonium cations and chlorine dopants heal defects in lead iodide perovskites. *Advanced Functional Materials*, 30, 1910004.

Nespurek, S. and Sworakowski, J. (1977). A differential method of analysis of steady-state space charge-limited current-voltage characteristics. *Physica Status Solidi A*, 41, 619–627.

Nespurek, S. and Sworakowski, J. (1980). Use of space charge-limited current measurements to determine the properties of energetic distributions of bulk traps. *Journal of Applied Physics*, 51, 2096–2102.

Nespurek, S., Zmescal, O., Sworakowski, J. (2008). Space charge-limited current in organic films: Some open problems. *Thin Solid Films*, 516, 8949–8962.

Neugebauer, S., Rauh, J., Deibel, C., Dyakonov, V. (2012). Investigation of electronic trap states in organic photovoltaic materials by current-based deep level transient spectroscopy. *Applied Physics Letters*, 100, 263304.

Nguyen, T.P. and Renaud, C. (2009). Interface effects on the defect state formation in organic devices. *Proceedings of SPIE XI*, 7213, 721314.

Nguyen, T.P. and Tran, V.H. (1995). Dielectric properties of poly(phenylene-vinylene) thin films. *Materials Science and Engineering B*, 31, 255–260.

Nguyen, T.P., Tran, V.H., Massardier, V. (1993a). Electrical conduction in poly (phenylenevinylene thin films). *Journal of Physics: Condensed Matter*, 5, 6243–6252.

Nguyen, T.P., Massardier, V., Tran, V.H., Guyot, A. (1993b). Studies of the polymer-metal interface in metal-PPV-metal devices. *Synthetic Metals*, 55, 235–240.

Nguyen, T.P., Giffard, M., Molinié, P. (1994). Magnetic and conductivity studies of poly (vinylene disulfide). *Journal of Chemical Physics*, 100, 8340–8346.

Nguyen, T.P., Ip, J., Jolinat, P., Destruel, P. (2001). XPS and sputtering study of the Alq3/electrode interfaces in organic light emitting diodes. *Applied Surface Science*, 172, 75–83.

Nguyen, T.P., Le Rendu, P., Gaudin, O., Lee, A.J.T., Jackman, R.B., Huang, C.H. (2006). Defect states investigation in poly(2-methoxy,5-(2'ethyl-hexyloxy)-1,4-phenylene vinylene) (MEH-PPV). *Thin Solid Films*, 511–512, 338–341.

Nguyen, T.P., Renaud, C., Huang, C.H., Lo, C.N., Lee, C.W., Hsu, C.S. (2008). Effect of electrical operation on the defect states in organic semiconductors. *Journal of Material Science: Materials in Electronics*, 19, 92–95.

Nguyen, T.P., Renaud, C., Reisdorffer, F., Wang, L. (2012). Degradation of PCBM: P3HT organic photovoltaic cells and structure changes as determined by defect investigations. *Journal of Energy Photonics*, 2, 021013.

Nguyen, T.P., Girault, P., Renaud, C., Reisdorffer, F., Le Rendu, P., Wang, L. (2014). Effects of the negative electrode contact on the performance of poly(hexylthiophene):6,6-phenyl-C61-butyric acid methyl ester based organic solar cells. *Journal of Applied Physics*, 115, 012013.

Nguyen, K.V., Payne, M.M., Anthony, J.E., Lee, J.H., Song, E., Kang, B., Cho, K., Lee, W.H. (2016). Grain boundary induced bias instability in soluble acene-based thin-film transistors. *Scientific Reports*, 6, 33224.

Nicholas, K.H. and Woods, J. (1964). The evaluation of electron trapping parameters from conductivity glow curves in cadmium sulphide. *British Journal of Physics*, 15, 783–795.

Nicolai, H.T., Mandoc, M.M., Blom, P.W.M. (2011). Electron traps in semiconducting polymers: Exponential versus Gaussian trap distribution. *Physical Review B*, 83, 195204.

Nicolai, H.T., Kuik, M., Wetzelaer, G.A.H., de Boer, B., Campbell, C., Risko, C., Brédas, J.L., Blom, P.W.M. (2012). Unification of trap-limited electron transport in semiconducting polymers. *Nature Materials*, 11, 882–887.

Nikiforov, M.P., Lai, B., Chen, W., Chen, S., Schaller, R.D., Strzalka, J., Maserb, J., Darling, S.B. (2013). Detection and role of trace impurities in high-performance organic solar cells. *Energy & Environment Science*, 6, 1513–1520.

Nikitendo, V.R., Heil, H., von Seggern, H. (2003). Space charge limited current in regioregular poly-3-hexyl-thiophene. *Journal of Applied Physics*, 94, 2480–2485.

Niles, E.T., Roehling, J.D., Yamagata, H., Wise, A.J., Spano, F.C., Moulé, A.J., Grey, J.K. (2012). J-aggregate behavior in poly-3- hexylthiophene nanofibers. *Journal of Physical Chemistry Letters*, 3, 259–263.

Niu, Q., Rohloff, R., Wetzelaer, G.-J.A.H., Blom, P.W.M., Crăciun, N.I. (2018). Hole trap formation in polymer light-emitting diodes under current stress. *Nature Materials*, 17, 557–562.

Noh, Y.Y., Ghim, J., Kang, S.J., Baeg, K.J., Kim, D.Y., Yase, K. (2006). Effect of light irradiation on the characteristics of organic field-effect transistors. *Journal of Applied Physics*, 100, 094501.

Norrman, K., Larsen, N.B., Krebs, F.C. (2006). Lifetimes of organic photovoltaics: Combining chemical and physical characterisation techniques to study degradation mechanisms. *Solar Energy Materials & Solar Cells*, 90, 2793–2814.

Norrman, K., Madsen, M.V., Gevorgyan, S.A., Krebs, F.C. (2010). Degradation patterns in water and oxygen of an inverted polymer solar cell. *Journal of the American Chemical Society*, 132, 1688316892.

Ohring, M. and Kasprzak, L. (2015). *Reliability and Failure of Electronic Materials and Devices*. Academic Press-Elsevier, London.

Pacios, R., Chatten, A.J., Kawano, K., Durrant, J.R., Bradley, D.D.C., Nelson, J. (2006). Effects of photo-oxidation on the performance of poly[2-methoxy-5-(3′,7′-dimethyloctyloxy)-1,4-phenylene vinylene]:[6,6]-phenyl C61-butyric acid methyl ester solar cells. *Advanced Functional Materials*, 16, 2117–2126.

Pahner, P., Kleemann, H., Burtone, L., Tietze, M.L., Fischer, J., Leo, K., Lüssem, B. (2013). Pentacene schottky diodes studied by impedance spectroscopy: Doping properties and trap response. *Physical Review B*, 88, 195205.

Pang, H.S., Xu, H., Tang, C., Meng, L.K., Ding, Y., Cai, H.T., Xiao, J., Liu, R.L., Zeng, W.Q., Huang, W. (2019). Nature of defect states within amorphous NPB investigated through drive-level capacitance profiling. *Journal of Physical Chemistry C*, 123, 165–174.

Pankov, J.I. (1971). *Optical Processes in Semiconductor*. Dover Publications, New York.

Park, S. and Son, H.J. (2019). Intrinsic photo-degradation and mechanism of polymer solar cells: The crucial role of nonfullerene acceptors. *Journal of Materials Chemistry A*, 7, 25830–25837.

Parthasarathy, G., Shen, C., Kahn, A., Forrest, S.R. (2001). Lithium doping of semiconducting organic charge transport materials. *Journal of Applied Physics*, 89, 4986–4992.

Pasveer, W., Cottaar, J., Tanase, C., Coehoorn, R., Bobbert, P., Blom, P., de Leeuw, D., Michels, M. (2005). Unified description of charge-carrier mobilities in disordered semiconducting polymers. *Physical Review Letters*, 94, 206601.

Petrovic, M., Maksudov, T., Panagiotopoulos, A., Serpetzoglou, E., Konidakis, I., Stylianakis, M.M., Stratakis, E., Kymakis, E. (2019). Limitations of a polymer-based hole-transporting layer for application in planar inverted perovskite solar cells. *Nanoscale Advances*, 1, 3107–3118.

Pingel, P., Arvind, M., Kölln, L., Steyrleuthner, R., Kraffert, F., Behrends, J., Janiet, S., Neher, D. (2016). p-Type doping of poly(3-hexylthiophene) with the strong lewis acid tris(pentafluorophenyl) borane. *Advanced Electronic Materials*, 2, 1600204.

Podzorov, V., Menard, E., Borissov, A., Kiryukhin, V., Rogers, 1J.A., Gershenson, M.E. (2004). Intrinsic charge transport on the surface of organic semiconductors. *Physical Review Letters*, 93, 086602.

Pospisil, J., Zmescal, O., Nespurek, S., Kracovic, J., Weiter, M., Kovalenco, A. (2019). Desnsity of bulk trap states of hybrid lead halide perovskite single crystals: Temperature modulated space-charge limited current. *Scientific Reports*, 9, 3332.

Proskuryakov, Y.Y., Durose, K., Taele, B.M., Oelting, S. (2007). Impedance spectroscopy of unetched CdTe / CdS solar cells – Equivalent circuit analysis. *Journal of Applied Physics*, 102, 024504.

Qi, Y. (2011). Investigation of organic films by atomic force microscopy: Structural, nanotribological and electrical properties. *Surface Science Reports*, 66, 379–393.

Rajagopal, A.A. (1998). Kahn Photoemission spectroscopy investigation of magnesium–Alq3 interfaces. *Journal of Applied Physics*, 84, 355–358.

Randall, J.T. and Wilkins, M.H.F. (1945). The phosphorescence of various solids II. *Proceedings of the Royal Society*, 184, 390–407.

Rau, U. (2007). Reciprocity relation between photovoltaic quantum efficiency and electroluminescent emission of solar cells. *Physical Review B*, 76, 085303.

Ray, D. and Narasimhan, K.L. (2008). Measurement of deep states in hole doped organic semiconductors. *Journal of Applied Physics*, 103, 093711.

Ray, B., Baradwaj, A.G., Boudouris, B.W., Alam. M.A. (2014). Defect characterization in organic semiconductors by forward bias capacitance–voltage (FB-CV) analysis. *The Journal of Physical Chemistry C*, 118, 17461–17466.

Reese, M.O., Gevorgyan, S.A., Jorgensen, M., Bundgaard, E., Kurtz, S.R., Ginley, D.S., Olson, D.C., Lloyd, M.T., Moryillo, P., Katz, E.A. et al. (2011). Consensus stability testing protocols for organic photovoltaic materials and devices. *Solar Energy Materials & Solar Cells*, 95, 1253–1267.

Renaud, C., Huang, C.H., Lee, C.W., Le Rendu, P., Nguyen, T.P. (2008). Study of trap states in polyfluorene based devices by using TSC technique. *Thin Solid Films*, 516, 7209–7213.

Reshchikov, M.A. and Morkoç, H. (2005). Luminescence properties of defects in GaN. *Journal of Applied Physics*, 97, 061301.

Risteska, A., Bedolla, J., Northrup, J.E., Bergholz, W., Wagner, V., Knipp, D. (2016). Gap states in small molecule thin-film transistors. *Advanced Electronic Materials*, 2, 1500179.

Rivnay, J., Mannsfeld, S.C.B., Miller, C.E., Salleo, A., Toney, M.F. (2012). Quantitative determination of organic semiconductor microstructure from the molecular to device scale. *Chemical Review*, 112, 5488–5519.

Rizvi, S.M.H., Mantri, P., Mazhari, B. (2014). Trap signature in steady-state current-voltage characteristics of organic diode. *Journal of Applied Physics*, 115, 244502.

Roesch, R., Faber, T., von Hauff, E., Brown, T.M., Lira-Cantu, M., Hoppe, H. (2015). Procedures and practices for evaluating thin-film solar cell stability. *Advanced Energy Materials*, 5, 1501407.

Rohloff, R., Kotadiya, N.B., Craciun, N.I., Blom, P.W.M., Wetzelaer, G.A.H. (2017). Electron and hole transport in the organic small molecule α-NPD. *Applied Physics Letters*, 110, 073301..

Rojek, K., Schmechel, R., Benson, N. (2019). MIS–TSC: A combination of the thermally stimulated current method and a metal-insulator-semiconductor device for unipolar trap spectroscopy. *Applied Physics Letters*, 114, 152104.

Root, S.E., Savagatrup, S., Printz, A.D., Rodriquez, D., Lipomi, D.J. (2017). Mechanical properties of organic semiconductors for stretchable, highly flexible, and mechanically robust electronics. *Chemical Reviews*, 117, 6467–6499.

Rose, A. (1963). *Concepts in Photoconductivity and Allied Problems*. John Wiley & Sons, New York.

Rosenberg, J.W., Legodi, M.J., Rakita, Y., Cahen, D., Diale, M. (2017). Laplace current deep level transient spectroscopy measurements of defect states in methylammonium lead bromide single crystals. *Journal of Applied Physics*, 122, 145701.

Roy, V.A.L., Pode, R.B., Gundu Rao, T.K., Djurišic, A.B., Baldacchini, G. (2004). Optical and ESR studies of tris(8-hydroxyquinoline) aluminum. *Materials Science and Engineering B*, 106, 85–88.

Sakanoue, T. and Sirringhaus, H. (2010). Band-like temperature dependence of mobility in a solution-processed organic semiconductor. *Nature Materials*, 9, 736–740.

Sasabe, H. and Kido. J. (2013). Development of high performance OLEDs for general lighting. *Journal of Materials Chemistry C*, 1, 1699–1707.

Savva, A., Georgiou, E., Papazoglou, G., Chrusou, A.Z., Kapnisis, K., Choulis, S.A. (2015). Photovoltaic analysis of the effects of PEDOT:PSS-additives hole selective contacts on the efficiency and lifetime performance of inverted organic solar cells. *Solar Energy Materials & Solar Cells*, 132, 507–514.

Scheinert, S., Paasch, G., Schrodner, M., Sensfuss, H.K.R.S., Doll, T. (2002). Subthreshold characteristics of field effect transistors based on poly(3-dodecylthiophene) and an organic insulator. *Journal of Applied Physics*, 92, 330–337.

Scher, H. and Montroll, E.W. (1975). Anomalous transit-time dispersion in amorphous solids. *Physical Review B*, 12, 2455–2477.

Schmidt, T.D., Jäger, L., Noguchi, Y., Ishii, H., Brütting, W. (2015). Analyzing degradation effects of organic light-emitting diodes via transient optical and electrical measurements. *Journal of Applied Physics*, 117, 215502.

Scholz, S., Kondakov, D., Lüssem, B., Leo, K. (2015). Degradation mechanisms and reactions in organic light-emitting devices. *Chemical Reviews*, 115, 8449–8503.

Schön, J.H. and Batlogg, B. (2001). Trapping in organic field-effect transistors. *Journal of Applied Physics*, 89, 336–342.

Schulz, P., Edri, E., Kirmayer, S., Hodes, G., Cahen, D., Kahn, A. (2014). Interface energetics in organo-metal halide perovskite-based photovoltaic cells. *Energy & Environmental Science*, 7, 1377–1381.

Schwarz, R. (1998). Dispersive transport in disordered semiconductors. *Journal of Non-Crystalline Solids*, 227–230, 148–152.

Scott, J.C., Kaufman, J.H., Brock, P.J., DiPietro, R., Salem, J., Goitia, J.A. (1996). Degradation and failure of MEH-PPV light-emitting diodes. *Journal of Applied Physics*, 79, 2745–2751.

Sehati, P., Braun, S., Fahlman, M. (2013). Energy level alignment in Au/pentacene/PTCDA trilayer stacks. *Chemical Physics Letters*, 583, 38–41.

Seki, K., Hayashi, N., Oji, H., Ito, E., Ouchi, Y., Ishii, H. (2001). Electronic structure of organic/metal interfaces. *Thin Solid Films*, 393, 298–303.

Shao, Y., Xiao, Z., Bi, C., Yuan, Y., Huang, J. (2014). Origin and elimination of photocurrent hysteresis by fullerene passivation in CH3NH3PbI3 planar heterojunction solar cells. *Nature Communications*, 5, 5784.

Shao, Y., Fang, Y., Li, T., Wang, Q., Dong, Q., Deng, Y., Yuan, Y., Wei, H., Wang, M., Gruverman, A. et al. (2016). Grain boundary dominated ion migration in polycrystalline organic-inorganic halide Perovskite film. *Energy & Environmental Science*, 9, 1752–1759.

Sharma, A., Yadav, S., Kumar, P., Chaudhuri, S.R., Ghosh, S. (2013). Defect states and their energetic position and distribution in organic molecular semiconductors. *Applied Physics Letters*, 102, 143301.

Shi, D., Aldinolfi, V., Comin, R., Yuan, M., Alarousu, E., Buin, A., Chen, Y., Hoogland, S., Rothenberger, A., Katsiev, K. et al. (2015). Low trap-state density and long carrier diffusion in organolead trihalide perovskite single crystals. *Science*, 347, 519–522.

Shockley, W. and Read, W.T. (1952). Statistics of the recombination of holes and electrons. *Physical Review*, 87, 835–842.

Shrotriya, V. and Yang, Y. (2005). Capacitance–voltage characterization of polymer light-emitting diodes. *Journal of Applied Physics*, 97, 054504.

Sirringhaus, H. (2009). Reliability of organic field-effect transistors. *Advanced Materials*, 21, 3859–3873.

Snaith, H.J., Abate, A., Ball, J.M., Eperon, G.E., Leijtens, T., Noel, N.K., Stranks, S.D., Wang, J.T.W., Wojciechowski, K., Zhang, W. (2014). Anomalous hysteresis in perovskite solar cells. *Journal of Physical Chemistry Letters*, 5, 1511–1515.

So, F. and Kondakov, D. (2010). Degradation mechanisms in small-molecule and polymer. Organic light-emitting diodes, *Advanced Materials*, 22, 3762–3777.

Son, D.Y., Kim, S.G., Seo, J.Y., Lee, S.H., Shin, H., Lee, D., Park, N.G. (2018). Universal approach toward hysteresis-free perovskite solar cell via defect engineering. *Journal of the American Chemical Society*, 140, 1358–1364.

Song, W., Li, Z., So, S.K., Qiu, Y., Zhu, Y., Cao, L. (2001). Dynamic SIMS characterization of interface structure of Ag/Alq3/NPB/ITO model devices. *Surface and Interface Analysis*, 32, 102–105.

Spano, F.C. and Silva, C. (2014). H- and J-Aggregate behavior in polymeric semiconductors. *Annual Reviews of Physical Chemistry*, 65, 477–500.

Stallinga, P., Gomes, H.L., Rost, H., Holmes, A.B., Harrison, M.G., Friend, R.H. (2000). Electronic levels in MEH-PPV. *Synthetic Metals*, 111–112, 535–537.

Stankevych, A., Vakhnin, A., Andrienko, D., Paterson, L., Geno, J., Fishchuk, I., Bassler, H., Kohler, A., Kadashchuk, A. (2021). Density of states of OLED host materials from thermally stimulated luminescence. *Physical Review Applied*, 15, 044050.

Stassen, A.F., de Boer, R.W.I., Iosad, N.N., Morpurgo, A.F. (2004). Influence of the gate dielectric on the mobility of rubrene single-crystal field-effect transistors. *Applied Physics Letters*, 85, 3899–3901.

Steiger, J., Karg, S., Schmechel, R., von Seggern, H. (2001). Aging induced traps in organic semiconductors. *Synthetic Metals*, 122, 49–52.

Stesmans, A. (1989). The. Si identical to Si3 defect at various (111)Si/SiO2 and (111)Si/Si3N4 interfaces. *Semiconductor Science and Technology*, 4, 1000–1011.

Stranks, S.D. (2017). Nonradiative losses in metal halide perovskites. *ACS Energy Letters*, 2, 1515–525.

Stranks, S.D., Burlakov, V.M., Leijtens, T., Ball, J.M., Goriely, A., Snaith, H.J. (2014). Recombination kinetics in organic-inorganic perovskites: Excitons, free charge, and subgap states. *Physical Review Applied*, 2, 034007.

Sudandana, C.S. (1998). Techniques and applications of electron spin resonance. *Bulletin Materials Science*, 1, 1–70.

Sutherland, L.J., Weerasinghe, H.C., Simon, G.P. (2021). A review on emerging barrier materials and encapsulation strategies for flexible perovskite and organic photovoltaics. *Advanced Energy Materials*, 11, 2101383.

Sze, S.M. (1981). *Physics of Semiconductor Devices*. Wiley-Interscience, New York.

Tajima, H. and Yasui, M. (2011). Determination of trap density function based on measurements of photoinduced charge carrier extraction using linearly increasing voltage at low temperatures. *Journal of the Physical Society of Japan*, 80, 063705.

Tajima, H., Suzuki, T., Kimata, M. (2012). Direct determination of trap density function based on the photoinduced charge carrier extraction technique. *Organic Electronics*, 13, 2272–2280.

Tales, I.A. (1981). Trap spectroscopy by the fractional glow technique. *Physica Status Solidi (A)*, 66, 65–75.

Tamai, Y., Ohkita, H., Namatame, M., Marumoto, K., Shimomura, S., Yamanari, T., Ito, S. (2016). Light-induced degradation mechanism in Poly(3 hexylthiophene)/fullerene blend solar cells. *Advanced Energy Materials*, 6(11), 1600171.

Tang, C.W. (1986). Two-layer organic photovoltaic cell. *Applied Physics Letters*, 48, 183–185.

Tang, C.W. and Van Slyke, S.A. (1987). Organic electroluminescent diodes. *Applied Physics Letters*, 51, 51–53.

Tauc, J., Grigorovici, R., Vancu, A. (1966). Optical properties and electronic structure of amorphous germanium. *Physica Status Solidi*, 15, 627–637.

Tian, Y., Peter, M., Unger, E., Abdellah, M., Zheng, K., Pullerits, T., Yartsev, A., Sundstrom, V., Scheblykin, I.G. (2015). Mechanistic insights into perovskite photoluminescence enhancement: Light curing with oxygen can boost yield thousandfold. *Journal of Physical Chemistry Chemical Physics*, 17, 24978.

Tong, X., Wang, N., Slootsky, M., Yu, J., Forrest, S.R. (2013). Intrinsic burn-in efficiency loss of small-molecule organic photovoltaic cells due to exciton-induced trap formation. *Solar Energy Materials & Solar Cells*, 118, 116–123.

Tress, W. (2017). Metal halide perovskites as mixed electronic–ionic conductors: Challenges and opportunities from hysteresis to memristivity. *Journal of Physical Chemistry Letters*, 8, 3106–3114.

Tress, W., Marinova, N., Moehl, T., Zakeeruddin, S.M., Nazeeruddin, M.K., Grätzel, M. (2015). Understanding the rate-dependent J–V hysteresis, slow time component, and aging in CH3NH3PbI3 perovskite solar cells: The role of a compensated electric field. *Energy & Environmental Science*, 8, 995–1004.

Tsumura, A., Koezuka, H., Ando, T. (1986). Macromolecular electronic device: Field-effect transistor with a polythiophene thin film. *Applied Physics Letters*, 49, 1210–1212.

Tuladhar, S.M., Poplavskyy, D., Choulis, S.A., Durrant, J.R., Bradley, D.D.C., Nelson, J. (2005). Ambipolar charge transport in films of methanofullerene and poly(phenylenevinylene)/methano fullerene blends. *Advanced Functional Materials*, 15, 1171–1182.

Turak, A. (2013). Interfacial degradation in organic optoelectronics. *RSC Advances*, 3, 6188–6225.

Turak, A., Grozea, D., Feng, X.D., Lu, Z.H., Aziz, A., Hor, A.M. (2002). Metal/Alq3 interface structures. *Applied Physics Letters*, 81, 766–768.

Van Slyke, S.A., Chen, C.H., Tang, C.W. (1996). Organic electroluminescent devices with improved stability. *Applied Physics Letters*, 69, 2160–2162.

Veres, J., Ogier, S.D., Leeming, S.W., Cupertino, D.C., Khaffaf, S.D. (2003). Low-k insulators as the choice of dielectrics in organicfield-effect transistors. *Advanced Functional Materials*, 13, 199–204.

Vissenberg, M.C.J.M. and Matters, M. (1998). Theory of the field-effect mobility in amorphous organic transistors. *Physical Review B*, 57, 12964–12967.

Walter, T., Herberholz, R., Müller, C., Schock, H.W. (1996). Determination of defect distributions from admittance measurements and application to Cu(In,Ga)Se2 based heterojunctions. *Journal of Applied Physics*, 80, 4411–4420.

Wang, Z., Lou, Y., Naka, S., Okada, H. (2011). Bias and temperature dependent charge transport in solution-processed small molecular mixed single layer organic light emitting devices. *Applied Physics Letters*, 98, 063302.

Wang, L., Wang, L., Chen, C.J., Chen, K.C., Hao, Z., Luo, Y., Sun, C., Wu, M.C., Yu, J., Han, Y. et al. (2021). Green InGaN quantum dots breaking through efficiency and bandwidth bottlenecks of micro-LEDs. *Laser Photonics Review*, 15, 2000406.

Watkins, N.J., Mäkinen, A.J., Gao, Y., Uchida, M., Kafafi, Z.H. (2006). Direct observation of the evolution of occupied and unoccupied energy levels of two silole derivatives at their interfaces with magnesium. *Journal of Applied Physics*, 100, 103706.

Weiss, S. and Kassingm, R. (1988). Deep level transient Fourier spectroscopy (DLTFS) – A technique for the analysis of deep level properties. *Solid-State Electronics*, 31, 1733–1742.

Wen, X., Feng, Y., Huang, S., Huang, F., Cheng, Y.B., Green, M., Ho-Baillie, A. (2016). Defect trapping states and charge carrier recombination in organic–inorganic halide perovskites. *Journal of Materials Chemistry C*, 4, 793–800.

Werner, A.G., Blochwitz, J., Pfeiffer, M., Leo, K. (2001). Field dependence of thermally stimulated currents in Alq3. Field dependence of thermally stimulated current in Alq3. *Journal of Applied Physics*, 90, 123–125.

Wessels, B.W. (1976). Determination of deep levels in Cu-doped GaP using transient-current spectroscopy. *Journal of Applied Physics*, 47, 1131–1133.

Wong, K.W., Yip, H.L., Luo, Y., Wong, K.Y., Lau, W.M., Low, K.H., Chow, H.F., Gao, Z.Q., Yeung, W.L., Chang, C.C. (2002). Blocking reactions between indium-tin oxide and poly (3,4-ethylene dioxythiophene): Poly(styrene sulphonate) with a self-assembly monolayer. *Applied Physics Letters*, 80, 2788–2790.

Wrachien, N., Lago, N., Rizzo, A., D'Alpaos, R., Stefani, A., Turatti, G., Muccini, M., Meneghesso, G., Cester, A. (2015). Effects of thermal and electrical stress on DH4T-based organic thin-film-transistors with PMMA gate dielectrics. *Microelectronics Reliability*, 55, 1790–1794.

Wu, X., Tuan Trinh, M., Niesner, D., Zhu, H., Norman, Z., Owen, J.S., Yaffe, O., Kudisch, B.J., Zhu, X.Y. (2015). Trap states in lead iodide perovskites. *Journal of the American Chemical Society*, 137, 2089–2096.

Wunderlich, B. (1973). *Macromolecular Physics Volume 1: Crystal Structure. Morphology, Defects*. Academic Press, New York and London.

Yan, Y., She, X.J., Zhu, H., Wang, S.D. (2011). Origin of bias stress induced instability of contact resistance in organic thin film transistors. *Organic Electronics*, 12, 823–826.

Yang, Y.S., Kim, S.H., Lee, J.I., Chu, H.Y., Do, L.M., Lee, H., Oh, J., Zyung, T., Ryu, M.K., Jang, M.S. (2002). Deep-level defect characteristics in pentacene organic thin films. *Applied Physics Letters*, 80, 1595–1597.

Yang, L., Yan, L., You, W. (2013). Organic solar cells beyond one pair of donor – acceptor: Ternary blends and more. *Journal of Physical Chemistry Letters*, 4, 1802–1810.

Yang, D., Yang, R., Wang, K., Wu, C., Zhu, X., Feng, J., Ren, X., Fang, G., Priya, S., Liu, S. (2018). High efficiency planar-type perovskite solar cells with negligible hysteresis using EDTA-complexed SnO2. *Nature Communications*, 9, 3239.

Yao, Y., Dong, H., Liu, F., Russell, T.P., Hu, W. (2017). Approaching intra- and interchain charge transport of conjugated polymers facilely by topochemical polymerized single crystals. *Advanced Materials*, 29, 170125.

Yazdan Mehr, M., Bahrami, A., van Driel, W.D., Fan, X.J., Davis, J.L., Zhang, G.Q. (2020). Degradation of optical materials in solid-state lighting systems. *International Materials Reviews*, 65, 102–128.

Young, R.J. and Lovell, P.A. (2011). *Introduction to Polymers*. CRC Press, Boca Ratton.

Yuan, Y. and Huang, J. (2016). Ion migration in organometal trihalide perovskite and its impact on photovoltaic efficiency and stability. *Accounts of Chemical Research*, 49, 286–293.

Yuan, Y., Wang, Q., Shao, Y., Lu, H., Li, T., Gruverman, A., Huang, J. (2016). Electric-field-driven reversible conversion between methylammonium lead triiodide perovskites and lead iodide at elevated temperatures. *Advanced Energy Materials*, 6, 1501803.

Zhao, C.X., Mao, A.Y., Xu, G. (2014). Junction capacitance and donor-acceptor interface of organic photovoltaics. *Applied Physics Letters*, 105, 063302.

Zhao, J., Li, Y., Yang, G., Jiang, K., Lin, H., Ade, H., Ma, W., Yan, H. (2016). Efficient organic solar cells processed from hydrocarbon solvents. *Nature Energy*, 1, 15027.

Zhao, W., Li, S., Yao, H., Zhang, S., Zhang, Y., Yang, B., Hou, J. (2017). Molecular optimization enables over 13% efficiency in organic solar cells. *Journal of the American Chemical Society*, 139, 7148–7151.

Zhu, H., Jia, W., Tang, X., Qu, F., Xu, J., Zhao, X., Ma, C., Wang, Y., Ye, S., Xiong, Z. (2020). Trap-enhanced intersystem crossing in tris(8-hydroxyquinoline) aluminum-based organic light-emitting diodes via in situ heating. *Journal of Physical Chemistry C*, 124, 3218–3223.

Index

A, B

C, D

E, F

G, H

I, K

L, M

N, O, P

R, S

T, V

Other titles from

iSTE

in

Electronics Engineering

2022

METHAPETTYPARAMBU PURUSHOTHAMA Jayakrishnan, PERRET Etienne, VENA Arnaud
Non-Volatile CBRAM/MIM Switching Technology for Electronically Reconfigurable Passive Microwave Devices: Theory and Methods for Application in Rewritable Chipless RFID

2021

NGUYEN Thien-Phap
Organic Electronics 1: Materials and Physical Processes
Organic Electronics 2: Applications and Marketing

2020

CAPPY Alain
Neuro-inspired Information Processing

FOUQUET François
Noise in Radio-Frequency Electronics and its Measurement

MAHDOUM Ali
CAD of Circuits and Integrated Systems

2018

MURET Pierre
Fundamentals of Electronics 2: Continuous-time Signals and Systems
Fundamentals of Electronics 3: Discrete-time Signals and Systems, and Quantized Level Systems

2017

MURET Pierre
Fundamentals of Electronics 1: Electronic Components and Elementary Functions

BUCCI Davide
Analog Electronics for Measuring Systems

2016

BAUDRAND Henri, TITAOUINE Mohammed, RAVEU Nathalie
The Wave Concept in Electromagnetism and Circuits: Theory and Applications

FANET Hervé
Ultra Low Power Electronics and Adiabatic Solutions

NDJOUNTCHE Tertulien
Digital Electronics 1: Combinational Logic Circuits
Digital Electronics 2: Sequential and Arithmetic Logic Circuits
Digital Electronics 3: Finite-state Machines

2015

DURAFFOURG Laurent, ARCAMONE Julien
Nanoelectromechanical Systems

2014

APPRIOU Alain
Uncertainty Theories and Multisensor Data Fusion

CONSONNI Vincent, FEUILLET Guy
Wide Band Gap Semiconductor Nanowires 1: Low-Dimensionality Effects and Growth
Wide Band Gap Semiconductor Nanowires 2: Heterostructures and Optoelectronic Devices

GAUTIER Jean-Luc
Design of Microwave Active Devices

LACAZE Pierre Camille, LACROIX Jean-Christophe
Non-volatile Memories

TEMPLIER François
OLED Microdisplays: Technology and Applications

THOMAS Jean-Hugh, YAAKOUBI Nourdin
New Sensors and Processing Chain

2013

COSTA François, GAUTIER Cyrille, LABOURE Eric, REVOL Bertrand
Electromagnetic Compatibility in Power Electronics

KORDON Fabrice, HUGUES Jérôme, CANALS Agusti, DOHET Alain
Embedded Systems: Analysis and Modeling with SysML, UML and AADL

LE TIEC Yannick
Chemistry in Microelectronics

2012

BECHERRAWY Tamer
Electromagnetism: Maxwell Equations, Wave Propagation and Emission

LALAUZE René
Chemical Sensors and Biosensors

LE MENN Marc
Instrumentation and Metrology in Oceanography

SAGUET Pierre
Numerical Analysis in Electromagnetics: The TLM Method

2011

ALGANI Catherine, RUMELHARD Christian, BILLABERT Anne-Laure
Microwaves Photonic Links: Components and Circuits

BAUDRANT Annie
Silicon Technologies: Ion Implantation and Thermal Treatment

DEFAY Emmanuel
Integration of Ferroelectric and Piezoelectric Thin Films: Concepts and Applications for Microsystems

DEFAY Emmanuel
Ferroelectric Dielectrics Integrated on Silicon

BESNIER Philippe, DÉMOULIN Bernard
Electromagnetic Reverberation Chambers

LANDIS Stefan
Nano-lithography

2010

LANDIS Stefan
Lithography

PIETTE Bernard
VHF / UHF Filters and Multicouplers

2009

DE SALVO Barbara
Silicon Non-volatile Memories / Paths of Innovation

DECOSTER Didier, HARARI Joseph
Optoelectronic Sensors

FABRY Pierre, FOULETIER Jacques
Chemical and Biological Microsensors / Applications in Fluid Media

GAUTIER Jacques
Physics and Operation of Silicon Devices in Integrated Circuits

MOLITON André
Solid-State Physics for Electronics

PERRET Robert
Power Electronics Semiconductor Devices

SAGUET Pierre
Passive RF Integrated Circuits

2008

CHARRUAU Stéphane
Electromagnetism and Interconnections

2007

RIPKA Pavel, TIPEK Alois
Modern Sensors Handbook

Printed and bound by CPI Group (UK) Ltd, Croydon, CR0 4YY
16/09/2026
14963082-0005